BIRDS AND E
SHARP EYⳊ,
AND OTHER PAPERS

***

BY

JOHN BURROUGHS

WITH AN INTRODUCTION BY
MARY E. BURT

AND A BIOGRAPHICAL SKETCH

British Library Cataloguing-in-Publication Data
A catalogue record for this book is available from
the British Library

CONTENTS

JOHN BURROUGHS

John Burroughs was born on April 3 1837 in Catskill Mountains near Roxbury in Delaware County, New York, United States. As a child he played on the slopes of the Catskill Mountains and worked on the family farm. He was enthralled by the birds and other wildlife around him. Burroughs developed an interest in learning, but his father believed the rudimentary education given at the local school was enough, and refused to pay for the higher education that Burroughs desired. At seventeen he left home to earn the money needed for college by teaching at a school in Olive, New York. Between 1854 and 1856 he worked as a teacher whilst completing his studies. He continued to teach until 1863.

In 1857, Burroughs left his teaching position in Illinois to find employment near his hometown and that same year, he married the pious Ursula North (1836-1917). After five years of marital discord, Ursula concluded that her husband's sexual demands were immoral. She suggested a short separation to encourage him to value chastity. Their separation lasted until 1864, during which, Burroughs valued other female company. He remained unfaithful after their reunion. In 1901, he met Clara Barrus (1864-1931), a physician at a psychiatric hospital. She was half his age, but was the love of his life. She moved into his house after Ursula died in 1917.

Burroughs' first published essay was *Expression* (1860). In 1864, Burroughs began work as a clerk at the Treasury and eventually became a federal banker. He worked there until the 1880s, but continued writing and acquired an interest in the poetry of Walt Whitman (1819-1892). The pair met in 1863 and became friends. Whitman encouraged Burroughs to develop his nature writings, as well as his essays on philosophy and literature. In 1867, Burroughs published *Notes on Walt Whitman as Poet and Person* which was the first biography and critical work on Whitman, and was anonymously edited by Whitman before it was published. In 1871, Burroughs' first collection of nature essays, *Wake Robin*, was published.

Burroughs left Washington for New York in 1873 where he bought a fruit farm in West Park, New York and built his Riverby estate. In 1895, he bought additional land near Riverby and built an Adirondack style cabin named Slabsides. There, Burroughs wrote and entertained visitors. His famous friends included Theodore Roosevelt (1858-1919), Henry Ford (1863-1947) - who gave him an automobile, and Thomas Edison (1847-1931). In 1899, Burroughs accompanied E. H Harriman (1848-1909) on his expedition to Alaska and also travelled to the Grand Canyon and Yosemite with John Muir (1838-1914).

In 1903, after publishing an article, *Real and Sham Nature History*, Burroughs began a publicised debate known as the nature fakers controversy where he condemned certain writers for their absurd representation of wildlife. He also criticised the naturalistic animal stories genre. This disagreement lasted for four years and included many environmental and political figures.

Burroughs was best known for his writings on wildlife and rural life and his writing achievements were recognised by his election as a member of the American Academy of Arts and Letters. Some of his essays recalled his trips to the Catskills, for example, *The Heart of the Southern Catskills*, depicts the ascent of Slide Mountain. Other Catskills essays commented on fishing, hiking or rafting. He was an enthusiastic fly fisherman and contributed some notable fishing essays to angling literature, including *Speckled Trout* (1870). *The Complete Writings of John Burroughs* runs to twenty three volumes. *Wake Robin* was the first and the following volumes were published regularly with the final two, *Under the Maples* (1921) and *The Last Harvest* (1922), being published posthumously by Clara Barrus. Burroughs also published a biography of *John James Audubon* (1902), *Camping and Tramping with Roosevelt* (1906), and *Bird and Bough* (1906). Shortly before his death, Burroughs suffered from lapses in memory and a decline in heart function. In February 1921, he had an operation to remove an abscess from his chest, after which his health worsened. He died on March 29 1921 on a train near Kingsville, Ohio. He was buried in Roxbury, New York on what would have been his eighty fourth birthday, at the foot of a rock he had termed Boyhood Rock.

BIOGRAPHICAL SKETCH.

Nature chose the spring of the year for the time of John Burroughs's birth. A little before the day when the wake-robin shows itself, that the observer might be on hand for the sight, he was born in Roxbury, Delaware County, New York, on the western borders of the Catskill Mountains; the precise date was April 3, 1837. Until 1863 he remained in the country about his native place, working on his father's farm, getting his schooling in the district school and neighboring academies, and taking his turn also as teacher. As he himself has hinted, the originality, freshness, and wholesomeness of his writings are probably due in great measure to the unliterary surroundings of his early life, which allowed his mind to form itself on unconventional lines, and to the later companionships with unlettered men, which kept him in touch with the sturdy simplicities of life.

From the very beginnings of his taste for literature, the essay was his favorite form. Dr. Johnson was the prophet of his youth, but he soon transferred his allegiance to Emerson, who for many years remained his "master enchanter." To cure himself of too close an imitation of the Concord seer, which showed itself in his first magazine article, Expression, he took to writing his sketches of nature, and about this time he fell in with the writings of Thoreau, which doubtless confirmed and encouraged him in this direction. But of all authors and of all men, Walt Whitman, in his personality and as a literary force, seems to have made the profoundest impression upon Mr. Burroughs, though doubtless Emerson had a greater influence on his style of writing.

Expression appeared in The Atlantic Monthly in 1860, and most of his contributions to literature have been in the form of papers first published in the magazines, and afterwards collected into books. He more than once paid tribute to his teachers in literature. His first book, now out of print, was Notes on Walt Whitman, as Poet and Person, published in 1867; and Whitman: A Study, which appeared in 1896, is a more extended treatment of the man and his poetry and philosophy. Birds and Poets, too, contains a paper on Whitman, entitled The Flight of the Eagle, besides an essay on Emerson, whom he also treated incidentally in his paper, Matthew Arnold on Emerson and Carlyle, in Indoor Studies; and the latter volume contains his essay on Thoreau.

In the autumn of 1863 he went to Washington, and in the following January entered the Treasury Department. He was for some years an assistant in the office of the Comptroller of the Currency, and later chief of the organization division of that Bureau. For some time he was keeper of one of the vaults, and for a great part of the day his only duty was to be at his desk. In these leisure hours his mind traveled off into the country, where his previous life had been spent, and with the help of his pen, always a faithful friend and magician, he lived over again those happy days, now happier still with the glamour of all past pleasures. In this way he wrote Wake-Robin and a part of Winter Sunshine. It must not be supposed, however, that he was deprived of outdoor pleasures while at Washington. On the contrary, he enjoyed many walks in the suburbs of the capital, and in those days the real country came up to the very edges of the city. His Spring at the Capital, Winter Sunshine, A March Chronicle, and other papers bear the fruit of his life on the Potomac. He went to England in 1871 on business for the Treasury Department, and again on his own account a dozen years later. The record of the two visits is to be found mainly in his chapters on An October Abroad, contained in the volume Winter Sunshine, and in the papers gathered into the volume Fresh Fields.

He resigned his place in the Treasury in 1873, and was appointed receiver of a broken national bank. Later, until 1885, his business occupation was that of a National Bank Examiner. An article contributed by him to The Century Magazine for March, 1881, on Broken Banks and Lax Directors, is perhaps the only literary outcome of this occupation, but the keen powers of observation, trained in the field of nature, could not fail to disclose themselves in analyzing columns of figures. After leaving Washington Mr. Burroughs bought a fruit farm at West Park, near Esopus, on the Hudson, and there building his house from the stones found in his fields, has given himself the best conditions for that humanizing of nature which constitutes the charm of his books. He was married in 1857 to a lady living in the New York village where he was at the time teaching. He keeps his country home the year round, only occasionally visiting New York. The cultivation of grapes absorbs the greater part of his time; but he has by no means given over letters. His work, which has long found ready acceptance both at home and abroad, is now passing into that security of fame which comes from its entrance into the school-life of American children.

Besides his outdoor sketches and the other papers already mentioned, Mr. Burroughs has written a number of critical essays on life and literature, published in Indoor Studies, and other volumes. He has also taken his readers into his confidence in An Egotistical Chapter, the final one of his Indoor Studies; and in the Introduction to the Riverside Edition of his writings he has given us further glimpses of his private intellectual life.

Probably no other American writer has a greater sympathy with, and a keener enjoyment of, country life in all its phases— farming, camping, fishing, walking—than has John Burroughs. His books are redolent of the soil, and have such "freshness and primal sweetness," that we need not be told that the pleasure he gets from his walks and excursions is by no means over when he steps inside his doors again. As he tells us on more than one occasion, he finds he can get much more out of his outdoor experiences by thinking them over, and writing them out afterwards.

Numbers 28, 36, and 92 of the Riverside Literature Series consist of selections from Mr. Burroughs's books. No. 28, which is entitled Birds and Bees, is made up of Bird Enemies and The

Tragedies of the Nests from the volume Signs and Seasons, An Idyl of the Honey-Bee from Pepacton, and The Pastoral Bees from Locusts and Wild Honey. The Introduction, by Miss Mary E. Burt, gives an account of the use of Mr. Burroughs's writings in Chicago schools.

In No. 36, Sharp Eyes, and Other Papers, the initial paper, Sharp Eyes, is drawn from Locusts and Wild Honey, The Apple comes from Winter Sunshine, A Taste of Maine Birch and Winter Neighbors from Signs and Seasons, and Notes by the Way (on muskrats, squirrels, foxes, and woodchucks) from Pepacton.

The collection called A Bunch of Herbs, and Other Papers, forming No. 92 of the Series, was designed with special reference to what the author has to say of trees and flowers, and contains A Bunch of Herbs from Pepacton, Strawberries from Locusts and Wild Honey, A March Chronicle and Autumn Tides from Winter Sunshine, A Spray of Pine and A Spring Relish from Signs and Seasons, and English Woods: A Contrast from Fresh Fields.

INTRODUCTION.

It is seldom that I find a book so far above children that I cannot share its best thought with them. So when I first took up one of John Burroughs's essays, I at once foresaw many a ramble with my pupils through the enchanted country that is found within its breezy pages. To read John Burroughs is to live in the woods and fields, and to associate intimately with all their little timid inhabitants; to learn that—

> "God made all the creatures and gave them our love and our fear, To give sign, we and they are his children, one family here."

When I came to use Pepacton in my class of the sixth grade, I soon found, not only that the children read better but that they came rapidly to a better appreciation of the finer bits of literature in their regular readers, while their interest in their new author grew quickly to an enthusiasm. Never was a little brother or sister more real to them than was "Peggy Mel" as she rushed into the hive laden with stolen honey, while her neighbors gossiped about it, or the stately elm that played sly tricks, or the log which proved to be a good bedfellow because it did not grumble. Burroughs's way of investing beasts, birds, insects, and inanimate things with human motives is very pleasing to children. They like to trace analogies between the human and the irrational, to think of a weed as a tramp stealing rides, of Nature as a tell-tale when taken by surprise.

The quiet enthusiasm of John Burroughs's essays is much healthier than the over-wrought dramatic action which sets all the nerves a-quiver,—nerves already stimulated to excess by the comedies and tragedies forced upon the daily lives of children. It is especially true of children living in crowded cities, shut away from the woods and hills, constant witnesses of the effects of human passion, that they need the tonic of a quiet literature rather than the stimulant of a stormy or dramatic one,—a literature which develops gentle feelings, deep thought, and a relish for what is homely and homespun, rather than a literature which calls forth excited feelings.

The essays in this volume are those in which my pupils have expressed an enthusiastic interest, or which, after careful reading, I have selected for future use. I have found in them few pages so hard as to require over much study, or a too frequent use of the dictionary. John Burroughs, more than almost any other writer of the time, has a prevailing taste for simple words and simple constructions. "He that runs may read" him. I have found many children under eleven years of age who could read a whole page without hesitating. If I discover some words which I foresee will cause difficulty, I place such on the blackboard and rapidly pronounce and explain them before the reading. Generally, however, I find the text the best interpreter of its words. What follows explains what goes before, if the child is led to read on to the end of the sentence. It is a mistake to allow children to be frightened away from choice reading by an occasional hard word. There is no better time than his reading lesson in which to teach a child that the hard things of life are to be grappled with and overcome. A mistake also, I think, is that toilsome process of explanation which I sometimes find teachers following, under the impression that it will be "parrot work" (as the stock phrase of the "institutes" has it) for the pupils to read anything which they do not clearly and fully comprehend. Teachers' definitions, in such cases, I have often noticed, are no better than dictionary definitions, and surely everybody knows that few more fruitless things than dictionary definitions are ever crammed into the memory of a child. Better far give free play to the native intelligence of the child, and trust it to apprehend, though it may not yet comprehend nor be able to express its apprehension in definition. On this subject I am glad to quote so high an authority as Sir Walter Scott: "Indeed I rather suspect that children derive impulses of a powerful and important kind from reading things which they do not comprehend, and therefore that to write down to children's understanding is a mistake. Set them on the scent and let them puzzle it out."

From time to time I have allowed my pupils to give me written reports from memory of these essays, and have often found these little compositions sparkling with pleasing information, or full of that childlike fun which is characteristic of the author. I have marked the errors in these exercises, and have given them back to the children to rewrite. Sometimes the second papers show careful correction-and sometimes the mistakes are partially neglected. Very often the child wishes to improve on the first composition, and so adds new blunders as well as creates new interest.

There is a law of self-preservation in Nature, which takes care of mistakes. Every human soul reaches toward the light in the most direct path open to it, and will correct its own errors as soon as it is developed far enough. There is no use in trying to force maturity; teachers who trouble children beyond all reason, and worry over their mistakes, are fumbling at the roots of young plants that will grow if they are let alone long enough.

The average mechanical work (spelling, construction of sentences, writing, etc.) is better under this method than when more time is devoted to the mechanics and less to the thought of composition. I have seen many reports of Burroughs's essays from the pens of children more pleasing and reliable than the essays of some professional reviewers; in these papers I often find the children adding little suggestions of their own; as, "Do birds dream?" One of the girls says her bird "jumps in its sleep." A little ten year old writes, "Weeds are unuseful flowers," and, "I like this book because there are real things in it." Another thinks she "will look more carefully" if she ever gets out into the country again. For the development of close observation and good feeling toward the common things of life, I know of no writings better than those of John Burroughs.

MARY E. BURT

JONES SCHOOL, CHICAGO, Sept. 1, 1887.

I. BIRDS & BEES

BIRDS.

BIRD ENEMIES.

How surely the birds know their enemies! See how the wrens and robins and bluebirds pursue and scold the cat, while they take little or no notice of the dog! Even the swallow will fight the cat, and, relying too confidently upon its powers of flight, sometimes swoops down so near to its enemy that it is caught by a sudden stroke of the cat's paw. The only case I know of in which our small birds fail to recognize their enemy is furnished by the shrike; apparently the little birds do not know that this modest-colored bird is an assassin. At least, I have never seen them scold or molest him, or utter any outcries at his presence, as they usually do at birds of prey. Probably it is because the shrike is a rare visitant, and is not found in this part of the country during the nesting season of our songsters.

But the birds have nearly all found out the trick the jay, and when he comes sneaking through the trees in May and June in quest of eggs, he is quickly exposed and roundly abused. It is amusing to see the robins hustle him out of the tree which holds their nest. They cry "Thief, thief!" to the top of their voices as they charge upon him, and the jay retorts in a voice scarcely less complimentary as he makes off.

The jays have their enemies also, and need to keep an eye on their own eggs. It would be interesting to know if jays ever rob jays, or crows plunder crows; or is there honor among thieves even in the feathered tribes? I suspect the jay is often punished by birds which are otherwise innocent of nest-robbing. One season I found a jay's nest in a small cedar on the side of a wooded ridge. It held five eggs, every one of which had been punctured. Apparently some bird had driven its sharp beak through their shells, with the sole intention of destroying them, for no part of the contents of the eggs had been removed. It looked like a case of revenge; as if some thrush or warbler, whose nest had suffered at the hands of the jays, had watched its opportunity, and had in this way retaliated upon its enemies. An egg for an egg. The jays were lingering near, very demure and silent, and probably ready to join a crusade against nest-robbers.

The great bugaboo of the birds is the owl. The owl snatches them from off their roosts at night, and gobbles up their eggs and young in their nests. He is a veritable ogre to them, and his presence fills them with consternation and alarm.

One season, to protect my early cherries I placed a large stuffed owl amid the branches of the tree. Such a racket as there instantly began about my grounds is not pleasant to think upon! The orioles and robins fairly "shrieked out their affright." The news instantly spread in every direction, and apparently every bird in town came to see that owl in the cherry-tree, and every bird took a cherry, so that I lost more fruit than if I had left the owl in-doors. With craning necks and horrified looks the birds alighted upon the branches, and between their screams would snatch off a cherry, as if the act was some relief to their outraged feelings.

The chirp and chatter of the young of birds which build in concealed or inclosed places, like the woodpeckers, the house wren, the high-hole, the oriole, is in marked contrast to the silence of the fledglings of most birds that build open and exposed nests. The young of the sparrows,—unless the social sparrow be an exception,—warblers, fly-catchers, thrushes, never allow a sound to escape them; and on the alarm note of their parents being heard, sit especially close and motionless, while the young of chimney swallows, woodpeckers, and orioles are very noisy. The latter, in its deep pouch, is quite safe from birds of prey, except perhaps the owl. The owl, I suspect, thrusts its leg into the cavities of woodpeckers and into the pocket-like nest of the oriole, and clutches and brings forth the birds in its talons. In one case which I heard of, a screech-owl had thrust its claw into a cavity in a tree, and grasped the head of a red-headed woodpecker; being apparently unable to draw its prey forth, it had thrust its own round head into the hole, and in some way became fixed there, and had thus died with the woodpecker in its talons.

The life of birds is beset with dangers and mishaps of which we know little. One day, in my walk, I came upon a goldfinch with the tip of one wing securely fastened to the feathers of its rump, by what appeared to be the silk of some caterpillar. The bird, though uninjured, was completely crippled, and could not fly a stroke. Its little body was hot and panting in my hands, as I carefully broke the fetter. Then it darted swiftly away with a happy cry. A record of all the accidents and tragedies of bird life for a single season would show many curious incidents. A friend of mine opened his box-stove one fall to kindle a fire in it, when he beheld in the black interior the desiccated forms of two bluebirds. The birds had probably taken refuge in the chimney during some cold spring storm, and had come down the pipe to the stove, from whence they were unable to ascend. A peculiarly touching little incident of bird life occurred to a caged female canary. Though unmated, it laid some eggs, and the happy bird was so carried away by her feelings that she would offer food to the eggs, and chatter and twitter, trying, as it seemed, to encourage them to eat! The incident is hardly tragic, neither is it comic.

Certain birds nest in the vicinity of our houses and outbuildings, or even in and upon them, for protection from their enemies, but they often thus expose themselves to a plague of the most deadly character.

I refer to the vermin with which their nests often swarm, and which kill the young before they are fledged. In a state of nature this probably never happens; at least I have never seen or heard of it happening to nests placed in trees or under rocks. It is the curse of civilization falling upon the birds which come too near man. The vermin, or the germ of the vermin, is probably conveyed to the nest in hen's feathers, or in straws and hairs picked up about the barn or hen-house. A robin's nest upon your porch or in your summer-house will occasionally become an intolerable nuisance from the swarms upon swarms of minute vermin with which it is filled. The parent birds stem the tide as long as they can, but are often compelled to leave the young to their terrible fate.

One season a phoebe-bird built on a projecting stone under the eaves of the house, and all appeared to go well till the young were nearly fledged, when the nest suddenly became a bit of purgatory. The birds kept their places in their burning bed till they could hold no longer, when they leaped forth and fell dead upon the ground.

After a delay of a week or more, during which I imagine the parent birds purified themselves by every means known to them, the couple built another nest a few yards from the first, and proceeded to rear a second brood; but the new nest developed into the same bed of torment that the first did, and the three young birds, nearly ready to fly, perished as they sat within it. The parent birds then left the place as if it had been accursed.

I imagine the smaller birds have an enemy in our native white-footed mouse, though I have not proof enough to convict him. But one season the nest of a chickadee which I was observing was broken up in a position where nothing but a mouse could have reached it. The bird had chosen a cavity in the limb of an apple-tree which stood but a few yards from the house. The cavity was deep, and the entrance to it, which was ten feet from the ground, was small. Barely light enough was admitted, when the sun was in the most favorable position, to enable one to make out the number of eggs, which was six, at the bottom of the dim interior. While one was peering in and trying to get his head out of his own light, the bird would startle him by a queer kind of puffing sound. She would not leave her nest like most birds, but really tried to blow or scare the intruder away; and after repeated experiments I could hardly refrain from jerking my head back when that little explosion of sound came up from the dark interior. One night, when incubation was about half finished, the nest was harried. A slight trace of hair or fur at the entrance led me to infer that some small animal was the robber. A weasel might have done it, as they sometimes climb trees, but I doubt if either a squirrel or a rat could have passed the entrance.

Probably few persons have ever suspected the cat-bird of being an egg-sucker; I do not know that she has ever been accused of such a thing, but there is something uncanny and disagreeable about her, which I at once understood, when I one day caught her in the very act of going through a nest of eggs.

A pair of the least fly-catchers, the bird which says chebec, chebec, and is a small edition of the pewee, one season built their nest where I had them for many hours each day under my observation. The nest was a very snug and compact structure placed in the forks of a small maple about twelve feet from the ground. The season before, a red squirrel had harried the nest of a wood-thrush in this same tree, and I was apprehensive that he would serve the fly-catchers the same trick; so, as I sat with my book in a summer-house near by, I kept my loaded gun within easy reach. One egg was laid, and the next morning, as I made my daily inspection of the nest, only a fragment of its empty shell was to be found. This I removed, mentally imprecating the rogue of a red squirrel. The birds were much disturbed by the event, but did not desert the nest, as I had feared they would, but after much inspection of it and many consultations together, concluded, it seems, to try again. Two more eggs were laid, when one day I heard the birds utter a sharp cry, and on looking up I saw a cat-bird perched upon the rim of the nest, hastily devouring the eggs. I soon regretted my precipitation in killing her, because such interference is generally unwise. It turned out that she had a nest of her own with five eggs in a spruce-tree near my window.

Then this pair of little fly-catchers did what I had never seen birds do before; they pulled the nest to pieces and rebuilt it in a peach-tree not many rods away, where a brood was successfully reared. The nest was here exposed to the direct rays of the noon-day sun, and to shield her young when the heat was greatest, the mother-bird would stand above them with wings slightly spread, as other birds have been know to do under like circumstances.

To what extent the cat-bird is a nest-robber I have no evidence, but that feline mew of hers, and that flirting, flexible tail, suggest something not entirely bird-like.

Probably the darkest tragedy of the nest is enacted when a snake plunders it. All birds and animals, so far I have observed, behave in a peculiar manner toward a snake. They seem to feel something of the loathing toward it that the human species experiences. The bark of a dog when he encounters a snake is different from that which he gives out on any other occasion; it is a mingled note of alarm, inquiry, and disgust.

One day a tragedy was enacted a few yards from where I was sitting with a book; two song-sparrows trying to defend their nest against a black snake. The curious, interrogating note of a chicken who had suddenly come upon the scene in his walk caused me to look up from my reading. There were the sparrows, with wings raised in a way peculiarly expressive of horror and dismay, rushing about a low clump of grass and bushes. Then, looking more closely, I saw the glistening form of the black snake and the quick movement of his head as he tried to seize the birds. The sparrows darted about and through the grass and weeds, trying to beat the snake off. Their tails and wings were spread, and, panting with the heat and the desperate struggle, they presented a most singular spectacle. They uttered no cry, not a sound escaped them; they were plainly speechless with horror and dismay. Not once did they drop their wings, and the peculiar expression of those uplifted palms, as it were, I shall never forget.

It occurred to me that perhaps here was a case of attempted bird-charming on the part of the snake, so I looked on from behind the fence. The birds charged the snake and harassed him from every side, but were evidently under no spell save that of courage in defending their nest. Every moment or two I could see the head and neck of the serpent make a sweep at the birds, when the one struck at would fall back, and the other would renew the assault from the rear. There appeared to be little danger that the snake could strike and hold one of the birds, though I trembled for them, they were so bold and approached so near to the snake's head. Time and again he sprang at them, but without success. How the poor things panted, and held up their wings appealingly! Then the snake glided off to the near fence, barely escaping the stone which I hurled at him. I found the nest rifled and deranged; whether it had contained eggs or young I know not. The male sparrow had cheered me many a day with his song, and I blamed myself for not having rushed at once to the rescue, when the arch enemy was upon him. There is probably little truth in the popular notion that snakes charm birds. The black snake is the most subtle, alert, and devilish of our snakes, and I have never seen him have any but young, helpless birds in his mouth.

We have one parasitical bird, the cow-bird, so-called because it walks about amid the grazing cattle and seizes the insects which their heavy tread sets going, which is an enemy of most of the smaller birds. It drops its egg in the nest of the song-sparrow, the social sparrow, the snow-bird, the vireos, and the wood-warblers, and as a rule it is the only egg in the nest that issues successfully. Either the eggs of the rightful owner of the nest are not hatched, or else the young are overridden and overreached by the parasite and perish prematurely.

Among the worst enemies of our birds are the so-called "collectors," men who plunder nests and murder their owners in the name of science. Not the genuine ornithologist, for no one is more careful of squandering bird life than he; but the sham ornithologist, the man whose vanity or affectation happens to take an ornithological turn. He is seized with an itching for a collection of eggs and birds because it happens to be the fashion, or because it gives him the air of a man of science. But in the majority of cases the motive is a mercenary one; the collector expects to sell these spoils of the groves and orchards. Robbing the nests and killing birds becomes a business with him. He goes about it systematically, and becomes expert in circumventing and slaying our songsters. Every town of any considerable size is infested with one or more of these bird highwaymen, and every nest in the country round about that the wretches can lay hands on is harried. Their professional term for a nest of eggs is "a clutch," a word that well expresses the work of their grasping, murderous fingers. They clutch and destroy in the germ the life and music of the woodlands. Certain of our natural history journals are mainly organs of communication between these human weasels. They record their exploits at nest-robbing and bird-slaying in their columns. One collector tells with gusto how he "worked his way" through an orchard, ransacking every tree, and leaving, as he believed, not one nest behind him. He had better not be caught working his way through my orchard. Another gloats over the number of Connecticut warblers—a rare bird—he killed in one season in Massachusetts. Another tells how a mocking-bird appeared in southern New England and was hunted down by himself and friend, its eggs "clutched," and the bird killed. Who knows how much the bird lovers of New England lost by that foul deed? The progeny of the birds would probably have returned to Connecticut to breed, and their progeny, or a part of them, the same, till in time the famous songster would have become a regular visitant to New England. In the same journal still another collector describes minutely how he outwitted three humming birds and captured their nests and eggs,—a clutch he was very proud of. A Massachusetts bird harrier boasts of his clutch of the eggs of that dainty little warbler, the blue yellow-back. One season he took two sets, the next five sets, the next four sets, besides some single eggs, and the next season four sets, and says he might have found more had he had more time. One season he took, in about twenty days, three from one tree. I have heard of a collector who boasted of having taken one hundred sets of the eggs of the marsh wren, in a single day; of another, who took in the same time, thirty nests of the yellow-breasted chat; and of still another, who claimed to have taken one thousand sets of eggs of different birds in one season. A large business has grown up under the influence of this collecting craze. One dealer in eggs has those of over five hundred species. He says that his business in 1883 was twice that of 1882; in 1884 it was twice that of 1883, and so on. Collectors vie with each other in the extent and variety of their cabinets. They not only obtain eggs in sets, but aim to have a number of sets of the same bird so as to show all possible variations. I hear of a private collection that contains twelve sets of kingbirds' eggs, eight sets of house-wrens' eggs, four sets mocking-birds' eggs, etc.; sets of eggs taken in low trees, high trees, medium trees; spotted sets, dark sets, plain sets, and light sets of the same species of bird. Many collections are made on this latter plan.

Thus are our birds hunted and cut off and all in the name of science; as if science had not long ago finished with these birds. She has weighed and measured, and dissected, and described them, and their nests, and eggs, and placed them in her cabinet; and the interest of science and of humanity now demands that this wholesale nest-robbing cease. These incidents I have given above, it is true, are but drops in the bucket, but the bucket would be more than full if we could get all the facts. Where one man publishes his notes, hundreds, perhaps thousands, say nothing, but go as silently about their nest-robbing as weasels.

It is true that the student of ornithology often feels compelled to take bird-life. It is not an easy matter to "name all the birds without a gun," though an opera-glass will often render identification entirely certain, and leave the songster unharmed; but once having mastered the birds, the true ornithologist leaves his gun at home. This view of the case may not be agreeable to that desiccated mortal called the "closet naturalist," but for my own part the closet naturalist is a person with whom I have very little sympathy. He is about the most wearisome and profitless creature in existence. With his piles of skins, his cases of eggs, his laborious feather-splitting, and his outlandish nomenclature, he is not only the enemy of the birds but the enemy of all those who would know them rightly.

Not the collectors alone are to blame for the diminishing numbers of our wild birds, but a large share of the responsibility rests upon quite a different class of persons, namely, the milliners. False taste in dress is as destructive to our feathered friends as are false aims in science. It is said that the traffic in the skins of our brighter plumaged birds, arising from their use by the milliners, reaches to hundreds of thousands annually. I am told of one middleman who collected from the shooters in one district, in four months, seventy thousand skins. It is a barbarous taste that craves this kind of ornamentation. Think of a woman or girl of real refinement appearing upon the street with her head gear adorned with the scalps of our songsters!

It is probably true that the number of our birds destroyed by man is but a small percentage of the number cut off by their natural enemies; but it is to be remembered that those he destroys are in addition to those thus cut off, and that it is this extra or artificial destruction that disturbs the balance of nature. The operation of natural causes keeps the birds in check, but the greed of the collectors and milliners tends to their extinction.

I can pardon a man who wishes to make a collection of eggs and birds for his own private use, if he will content himself with one or two specimens of a kind, though he will find any collection much less satisfactory and less valuable than he imagines, but the professional nest-robber and skin collector should be put down, either by legislation or with dogs and shotguns.

I have remarked above that there is probably very little truth in the popular notion that snakes can "charm" birds. But two of my correspondents have each furnished me with an incident from his own experience, which seems to confirm the popular belief. One of them writes from Georgia as follows:—

"Some twenty-eight years ago I was in Calaveras County, California, engaged in cutting lumber. One day in coming out of the camp or cabin, my attention was attracted to the curious action of a quail in the air, which, instead of flying low and straight ahead as usual, was some fifty feet high, flying in a circle, and uttering cries of distress. I watched the bird and saw it gradually descend, and following with my eye in a line from the bird to the ground saw a large snake with head erect and some ten or twelve inches above the ground, and mouth wide open, and as far as I could see, gazing intently on the quail (I was about thirty feet from the snake). The quail gradually descended, its circles growing smaller and smaller and all the time uttering cries of distress, until its feet were within two or three inches of the mouth of the snake; when I threw a stone, and though not hitting the snake, yet struck the ground so near as to frighten him, and he gradually started off. The quail, however, fell to the ground, apparently lifeless. I went forward and picked it up and found it was thoroughly overcome with fright, its little heart beating as if it would burst through the skin. After holding it in my hand a few moments it flew away. I then tried to find the snake, but could not. I am unable to say whether the snake was venomous or belonged to the constricting family, like the black snake. I can well recollect it was large and moved off rather slow. As I had never seen anything of the kind before, it made a great impression on my mind, and after the lapse of so long a time, the incident appears as vivid to me as though it had occurred yesterday."

It is not probable that the snake had its mouth open; its darting tongue may have given that impression.

The other incident comes to me from Vermont. "While returning from church in 1876," says the writer, "as I was crossing a bridge... I noticed a striped snake in the act of charming a song-sparrow. They were both upon the sand beneath the bridge.

The snake kept his head swaying slowly from side to side, and darted his tongue out continually. The bird, not over a foot away, was facing the snake, hopping from one foot to the other, and uttering a dissatisfied little chirp. I watched them till the snake seized the bird, having gradually drawn nearer. As he seized it, I leaped over the side of the bridge; the snake glided away and I took up the bird, which he had dropped. It was too frightened to try to fly and I carried it nearly a mile before it flew from my open hand."

If these observers are quite sure of what they saw, then undoubtedly snakes have the power to draw birds within their grasp. I remember that my mother told me that while gathering wild strawberries she had on one occasion come upon a bird fluttering about the head of a snake as if held there by a spell. On her appearance, the snake lowered its head and made off, and the panting bird flew away. A neighbor of mine killed a black snake which had swallowed a full-grown red squirrel, probably captured by the same power of fascination.

THE TRAGEDIES OF THE NESTS

The life of the birds, especially of our migratory song-birds, is a series of adventures and of hair-breadth escapes by flood and field. Very few of them probably die a natural death, or even live out half their appointed days. The home instinct is strong in birds as it is in most creatures; and I am convinced that every spring a large number of those which have survived the Southern campaign return to their old haunts to breed. A Connecticut farmer took me out under his porch, one April day, and showed me a phoebe bird's nest six stories high. The same bird had no doubt returned year after year; and as there was room for only one nest upon her favorite shelf, she had each season reared a new superstructure upon the old as a foundation. I have heard of a white robin—an albino—that nested several years in succession in the suburbs of a Maryland city. A sparrow with a very marked peculiarity of song I have heard several seasons in my own locality. But the birds do not all live to return to their old haunts: the bobolinks and starlings run a gauntlet of fire from the Hudson to the Savannah, and the robins and meadow-larks and other song-birds are shot by boys and pot-hunters in great numbers,—to say nothing of their danger from hawks and owls. But of those that do return, what perils beset their nests, even in the most favored localities! The cabins of the early settlers, when the country was swarming with hostile Indians, were not surrounded by such dangers. The tender households of the birds are not only exposed to hostile Indians in the shape of cats and collectors, but to numerous murderous and bloodthirsty animals, against whom they have no defense but concealment. They lead the darkest kind of pioneer life, even in our gardens and orchards, and under the walls of our houses. Not a day or a night passes, from the time the eggs are laid till the young are flown, when the chances are not greatly in favor of the nest being rifled and its contents devoured,—by owls, skunks, minks, and coons at night, and by crows, jays, squirrels, weasels, snakes, and rats during the day. Infancy, we say, is hedged about by many perils; but the infancy of birds is cradled and pillowed in peril. An old Michigan settler told me that the first six children that were born to him died; malaria and teething invariably carried them off when they had reached a certain age; but other children were born, the country improved, and by and by the babies weathered the critical period and the next six lived and grew up. The birds, too, would no doubt persevere six times and twice six times, if the season were long enough, and finally rear their family, but the waning summer cuts them short, and but a few species have the heart and strength to make even the third trial.

The first nest-builders in spring, like the first settlers near hostile tribes, suffer the most casualties. A large portion of the nests of April and May are destroyed; their enemies have been many months without eggs and their appetites are keen for them. It is a time, too, when other food is scarce, and the crows and squirrels are hard put. But the second nests of June, and still more the nests of July and August, are seldom molested. It is rarely that the nest of the goldfinch or the cedar-bird is harried.

My neighborhood on the Hudson is perhaps exceptionally unfavorable as a breeding haunt for birds, owing to the abundance of fish-crows and of red squirrels; and the season of which this chapter is mainly a chronicle, the season of 1881, seems to have been a black-letter one even for this place, for at least nine nests out of every ten that I observed during that spring and summer failed of their proper issue. From the first nest I noted, which was that of a bluebird,—built (very imprudently I thought at the time) in a squirrel-hole in a decayed apple-tree, about the last of April, and which came to naught, even the mother-bird, I suspect, perishing by a violent death,—to the last, which was that of a snow-bird, observed in August, among the Catskills, deftly concealed in a mossy bank by the side of a road that skirted a wood, where the tall thimble blackberries grew in abundance, from which the last young one was taken, when it was about half grown, by some nocturnal walker or daylight prowler, some untoward fate seemed hovering about them. It was a season of calamities, of violent deaths, of pillage and massacre, among our feathered neighbors. For the first time I noticed that the orioles were not safe in their strong, pendent nests. Three broods were started in the apple-trees, only a few yards from the house, where, for previous seasons, the birds had nested without molestation; but this time the young were all destroyed when about half grown. Their chirping and chattering, which was so noticeable one day, suddenly ceased the next. The nests were probably plundered at night, and doubtless by the little red screech-owl, which I know is a denizen of these old orchards, living in the deeper cavities of the trees. The owl could alight on the top of the nest, and easily thrust his murderous claw down into its long pocket and seize the young and draw them forth. The tragedy of one of the nests was heightened, or at least made more palpable, by one of the half-fledged birds, either in its attempt to escape or while in the clutches of the enemy, being caught and entangled in one of the horse-hairs by which the nest was stayed and held to the limb above. There it hung bruised and dead, gibbeted to its own cradle. This nest was the theatre of another little tragedy later in the season. Some time in August a bluebird, indulging its propensity to peep and pry into holes and crevices, alighted upon it and probably inspected the interior; but by some unlucky move it got its wings entangled in this same fatal horse-hair. Its efforts to free itself appeared only to result in its being more securely and hopelessly bound; and there it perished; and there its form, dried and embalmed by the summer heats, was yet hanging in September, the outspread wings and plumage showing nearly as bright as in life.

A correspondent writes me that one of his orioles got entangled in a cord while building her nest, and that though by the aid of a ladder he reached and liberated her, she died soon afterward. He also found a "chippie" (called also "hair bird") suspended from a branch by a horse-hair, beneath a partly constructed nest. I heard of a cedar-bird caught and destroyed in the same way, and of two young bluebirds, around whose legs a horse-hair had become so tightly wound that the legs withered up and dropped off. The birds became fledged, and left the nest with the others. Such tragedies are probably quite common.

Before the advent of civilization in this country, the oriole probably built a much deeper nest than it usually does at present. When now it builds in remote trees and along the borders of the woods, its nest, I have noticed, is long and gourd-shaped; but in orchards and near dwellings it is only a deep cup or pouch. It shortens it up in proportion as the danger lessens. Probably a succession of disastrous years, like the one under review, would cause it to lengthen it again beyond the reach of owl's talons or jay-bird's beak.

The first song-sparrow's nest I observed in the spring of 1881 was in the field under a fragment of a board, the board being raised from the ground a couple of inches by two poles. It had its full complement of eggs, and probably sent forth a brood of young birds, though as to this I cannot speak positively, as I neglected to observe it further. It was well sheltered and concealed, and was not easily come at by any of its natural enemies, save snakes and weasels. But concealment often avails little. In May, a song-sparrow, that had evidently met with disaster earlier in the season, built its nest in a thick mass of woodbine against the side of my house, about fifteen feet from the ground. Perhaps it took the hint from its cousin, the English sparrow. The nest was admirably placed, protected from the storms by the overhanging eaves and from all eyes by the thick screen of leaves. Only by patiently watching the suspicious bird, as she lingered near with food in her beak, did I discover its whereabouts. That brood is safe, I thought, beyond doubt. But it was not; the nest was pillaged one night, either by an owl, or else by a rat that had climbed into the vine, seeking an entrance to the house. The mother-bird, after reflecting upon her ill-luck about a week, seemed to resolve to try a different system of tactics and to throw all appearances of concealment aside. She built a nest few yards from the house beside the drive, upon a smooth piece of greensward. There was not a weed or a shrub or anything whatever to conceal it or mark its site. The structure was completed and incubation had begun before I discovered what was going on. "Well, well," I said, looking down upon the bird almost at my feet, "this is going to the other extreme indeed; now, the cats will have you." The desperate little bird sat there day after day, looking like a brown leaf pressed down in the short green grass. As the weather grew hot, her position became very trying. It was no longer a question of keeping the eggs warm, but of keeping them from roasting. The sun had no mercy on her, and she fairly panted in the middle of the day. In such an emergency the male robin has been known to perch above the sitting female and shade her with his outstretched wings. But in this case there was no perch for the male bird, had he been disposed to make a sunshade of himself. I thought to lend a hand in this direction myself, and so stuck a leafy twig beside the nest. This was probably an unwise interference; it guided disaster to the spot; the nest was broken up, and the mother-bird was probably caught, as I never saw her afterward.

For several previous summers a pair of kingbirds had reared, unmolested, a brood of young in an apple-tree, only a few yards from the house; but during this season disaster overtook them also. The nest was completed, the eggs laid, and incubation had begun, when, one morning about sunrise, I heard cries of distress and alarm proceed from the old apple-tree. Looking out of the window I saw a crow, which I knew to be a fish-crow, perched upon the edge of the nest, hastily bolting the eggs. The parent birds, usually so ready for the attack, seemed over-come with grief and alarm. They fluttered about in the most helpless and bewildered manner, and it was not till the robber fled on my approach that they recovered themselves and charged upon him. The crow scurried away with upturned, threatening head, the furious kingbirds fairly upon his back. The pair lingered around their desecrated nest for several days, almost silent, and saddened by their loss, and then disappeared. They probably made another trial elsewhere.

The fish-crow only fishes when it has destroyed all the eggs and young birds it can find. It is the most despicable thief and robber among our feathered creatures. From May to August, it is gorged with the fledglings of the nest. It is fortunate that its range is so limited. In size it is smaller than the common crow, and is a much less noble and dignified bird. Its caw is weak and feminine—a sort of split and abortive caw, that stamps it the sneak-thief it is. This crow is common farther south, but is not found in this State, so far as I have observed, except in the valley of the Hudson.

One season a pair of them built a nest in a Norway Spruce that stood amid a dense growth of other ornamental trees near a large unoccupied house. They sat down amid plenty. The wolf established himself in the fold. The many birds—robins, thrushes, finches, vireos, pewees—that seek the vicinity of dwellings (especially of these large country residences with their many trees and park-like grounds), for the greater safety of their eggs and young, were the easy and convenient victims of these robbers. They plundered right and left, and were not disturbed till their young were nearly fledged, when some boys, who had long before marked them as their prize, rifled the nest.

The song-birds nearly all build low; their cradle is not upon the tree-top. It is only birds of prey that fear danger from below more than from above, and that seek the higher branches for their nests. A line five feet from the ground would run above more than half the nests, and one ten feet would bound more than three fourths of them. It is only the oriole and the wood pewee that, as a rule, go higher than this. The crows and jays and other enemies of the birds have learned to explore this belt pretty thoroughly. But the leaves and the protective coloring of most nests baffle them as effectually, no doubt as they do the professional oölogist. The nest of the red-eyed vireo is one of the most artfully placed in the wood. It is just beyond the point where the eye naturally pauses in its search; namely, on the extreme end of the lowest branch of the tree, usually four or five feet from the ground. One looks up and down through the tree,—shoots his eye-beams into it as he might discharge his gun at some game hidden there, but the drooping tip of that low horizontal branch—who would think of pointing his piece just there? If a crow or other marauder were to alight upon the branch or upon those above it, the nest would be screened from him by the large leaf that usually forms a canopy immediately above it. The nest-hunter standing at the foot of the tree and looking straight before him, might discover it easily, were it not for its soft, neutral gray tint which blends so thoroughly with the trunks and branches of trees. Indeed, I think there is no nest in the woods—no arboreal nest—so well concealed. The last one I saw was a pendent from the end of a low branch of a maple, that nearly grazed the clapboards of an unused hay-barn in a remote backwoods clearing. I peeped through a crack and saw the old birds feed the nearly fledged young within a few inches of my face. And yet the cow-bird finds this nest and drops her parasitical egg in it. Her tactics in this as in other cases are probably to watch the movements of the parent bird. She may often be seen searching anxiously through the trees or bushes for a suitable nest, yet she may still oftener be seen perched upon some good point of observation watching the birds as they come and go about her. There is no doubt that, in many cases, the cow-bird makes room for her own illegitimate egg in the nest by removing one of the bird's own. When the cow-bird finds two or more eggs in a nest in which she wishes to deposit her own, she will remove one of them. I found a sparrow's nest with two sparrow's eggs and one cow-bird's egg, another egg lying a foot or so below it on the ground. I replaced the ejected egg, and the next day found it again removed, and another cow-bird's egg in its place; I put it back the second time, when it was again ejected, or destroyed, for I failed to find it anywhere. Very alert and sensitive birds like the warblers often bury the strange egg beneath a second nest built on top of the old. A lady, living in the suburbs of an eastern city, one morning heard cries of distress from a pair of house-wrens that had a nest in a honeysuckle on her front porch. On looking out of the window, she beheld this little comedy—comedy from her point of view, but no doubt grim-tragedy from the point of view of the wrens; a cow-bird with a wren's egg in its beak running rapidly along the walk with the outraged wrens forming a procession behind it, screaming, scolding, and gesticulating as only these voluble little birds can. The cow-bird had probably been surprised in the act of violating the nest, and the wrens were giving her a piece of theirs minds.

Every cow-bird is reared at the expense of two or more song-birds. For every one of these dusky little pedestrians there amid the grazing cattle there are two more sparrows, or vireos, or warblers, the less. It is a big price to pay—two larks for a bunting-two sovereigns for a shilling; but Nature does not hesitate occasionally to contradict herself in just this way. The young of the cow-bird is disproportionately large and aggressive, one might say hoggish. When disturbed it will clasp the nest and scream, and snap its beak threateningly. One hatched out in a song-sparrow's nest which was under my observation, and would soon have overridden and overborne the young sparrow, which came out of the shell a few hours later, had I not interfered from time to time and lent the young sparrow a helping hand. Every day I would visit the nest and take the sparrow out from under the pot-bellied interloper and place it on top so that presently it was able to hold its own against its enemy. Both birds became fledged and left the nest about the same time. Whether the race was an even one after that, I know not.

I noted but two warblers' nests during that season, one of the black-throated blue-back and one of the redstart,—the latter built in an apple-tree but a few yards from a little rustic summer-house where I idle away many summer days. The lively little birds, darting and flashing about, attracted my attention for a week before I discovered their nest. They probably built it by working early in the morning, before I appeared upon the scene, as I never saw them with material in their beaks. Guessing from their movements that the nest was in a large maple that stood near by, I climbed the tree and explored it thoroughly, looking especially in the forks of the branches, as the authorities say these birds build in a fork. But no nest could I find. Indeed, how can one by searching find a bird's nest? I overshot the mark; the nest was much nearer me, almost under my very nose, and I discovered it, not by searching but by a casual glance of the eye, while thinking of other matters. The bird was just settling upon it as I looked up from my book and caught her in the act. The nest was built near the end of a long, knotty, horizontal branch of an apple-tree, but effectually hidden by the grouping of the leaves; it had three eggs, one of which proved to be barren. The two young birds grew apace, and were out of the nest early in the second week; but something caught one of them the first night. The other probably grew to maturity, as it disappeared from the vicinity with its parents after some days.

The blue-back's nest was scarcely a foot from the ground, in a little bush situated in a low, dense wood of hemlock and beech and maple, amid the Catskills,—a deep, massive, elaborate structure, in which the sitting bird sank till her beak and tail alone were visible above the brim. It was a misty, chilly day when I chanced to find the nest, and the mother-bird knew instinctively that it was not prudent to leave her four half incubated eggs uncovered and exposed for a moment. When I sat down near the nest she grew very uneasy, and after trying in vain to decoy me away by suddenly dropping from the branches and dragging herself over the ground as if mortally wounded, she approached and timidly and half doubtingly covered her eggs within two yards of where I sat. I disturbed her several times to note her ways. There came to be something almost appealing in her looks and manner, and she would keep her place on her precious eggs till my outstretched hand was within a few feet of her. Finally, I covered the cavity of the nest with a dry leaf. This she did not remove with her beak, but thrust her head deftly beneath it and shook it off upon the ground. Many of her sympathizing neighbors, attracted by her alarm note, came and had a peep at the intruder and then flew away, but the male bird did not appear upon the scene. The final history of this nest I am unable to give, as I did not again visit it till late in the season, when, of course, it was empty.

Years pass without my finding a brown-thrasher's nest; it is not a nest you are likely to stumble upon in your walk; it is hidden as a miser hides his gold, and watched as jealously. The male pours out his rich and triumphant song from the tallest tree he can find, and fairly challenges you to come and look for his treasures in his vicinity. But you will not find them if you go. The nest is somewhere on the outer circle of his song; he is never so imprudent as to take up his stand very near it. The artists who draw those cosy little pictures of a brooding mother-bird with the male perched but a yard away in full song, do not copy from nature. The thrasher's nest I found thirty or forty rods from the point where the male was wont to indulge in his brilliant recitative. It was in an open field under a low ground-juniper. My dog disturbed the sitting bird as I was passing near. The nest could be seen only by lifting up and parting away the branches. All the arts of concealment had been carefully studied. It was the last place you would think of looking, and, if you did look, nothing was visible but the dense green circle of the low-spreading juniper. When you approached, the bird would keep her place till you had begun to stir the branches, when she would start out, and, just skimming the ground, make a bright brown line to the near fence and bushes. I confidently expected that this nest would escape molestation, but it did not. Its discovery by myself and dog probably opened the door for ill luck, as one day, not long afterward, when I peeped in upon it, it was empty. The proud song of the male had ceased from his accustomed tree, and the pair were seen no more in that vicinity.

The phoebe-bird is a wise architect, and perhaps enjoys as great an immunity from danger, both in its person and its nest, as any other bird. Its modest, ashen-gray suit is the color of the rocks where it builds, and the moss of which it makes such free use gives to its nest the look of a natural growth or accretion. But when it comes into the barn or under the shed to build, as it so frequently does, the moss is rather out of place. Doubtless in time the bird will take the hint, and when she builds in such places will leave the moss out. I noted but two nests, the summer I am speaking of: one, in a barn, failed of issue, on account of the rats, I suspect, though the little owl may have been the depredator; the other, in the woods, sent forth three young. This latter nest was most charmingly and ingeniously placed. I discovered it while in quest of pond-lilies, in a long, deep level stretch of water in the woods. A large tree had blown over at the edge of the water, and its dense mass of up-turned roots, with the black, peaty soil filling the interstices, was like the fragment of a wall several feet high, rising from the edge of the languid current. In a niche in this earthy wall, and visible and accessible only from the water, a phoebe had built her nest, and reared her brood. I paddled my boat up and came alongside prepared to take the family aboard. The young, nearly ready to fly, were quite undisturbed by my presence, having probably been assured that no danger need be apprehended from that side. It was not a likely place for minks, or they would not have been so secure.

I noted but one nest of the wood pewee, and that, too, like so many other nests, failed of issue. It was saddled upon a small dry limb of a plane-tree that stood by the roadside, about forty feet from the ground. Every day for nearly a week, as I passed by I saw the sitting bird upon the nest. Then one morning she was not in her place, and on examination the nest proved to be empty— robbed, I had no doubt, by the red squirrels, as they were very abundant in its vicinity, and appeared to make a clean sweep of every nest. The wood pewee builds an exquisite nest, shaped and finished as if cast in a mould. It is modeled without and within with equal neatness and art, like the nest of the humming-bird and the little gray gnat-catcher. The material is much more refractory than that used by either of these birds, being, in the present case, dry, fine cedar twigs; but these were bound into a shape as rounded and compact as could be moulded out of the most plastic material. Indeed, the nest of this bird looks precisely like a large, lichen-covered, cup-shaped excrescence of the limb upon which it is placed. And the bird, while sitting, seems entirely at ease. Most birds seem to make very hard work of incubation. It is a kind of martyrdom which appears to tax all their powers of endurance. They have such a fixed, rigid, predetermined look, pressed down into the nest and as motionless as if made of cast-iron. But the wood pewee is an exception. She is largely visible above the rim of the nest. Her attitude is easy and graceful; she moves her head this way and that, and seems to take note of whatever goes on about her; and if her neighbor were to drop in for a little social chat, she could doubtless do her part. In fact, she makes light and easy work of what, to most other birds, is such a serious and engrossing matter. If it does not look like play with her, it at least looks like leisure and quiet contemplation.

There is no nest-builder that suffers more from crows and squirrels and other enemies than the wood-thrush. It builds as openly and unsuspiciously as if it thought the whole world as honest as itself. Its favorite place is the fork of a sapling, eight or ten feet from the ground, where it falls an easy prey to every nest-robber that comes prowling through the woods and groves. It is not a bird that skulks and hides, like the cat-bird, the brown-thrasher, the chat, or the cheewink, and its nest is not concealed with the same art as theirs. Our thrushes are all frank, open-mannered birds; but the veery and the hermit build upon the ground, where they at least escape the crows, owls, and jays, and stand a better chance to be overlooked, by the red squirrel and weasel also; while the robin seeks the protection of dwellings and out-buildings. For years I have not known the nest of a wood-thrush to succeed. During the season referred to I observed but two, both apparently a second attempt, as the season was well advanced, and both failures. In one case, the nest was placed in a branch that an apple tree, standing near a dwelling, held out over the highway. The structure was barely ten feet above the middle of the road, and would just escape a passing load of hay. It was made conspicuous by the use of a large fragment of newspaper in its foundation—an unsafe material to build upon in most cases. Whatever else the press may guard, this particular newspaper did not guard this nest from harm. It saw the egg and probably the chick, but not the fledgeling. A murderous deed was committed above the public highway, but whether in the open day or under cover of darkness I have no means of knowing. The frisky red squirrel was doubtless the culprit. The other nest was in a maple sapling, within a few yards of the little rustic summer-house already referred to. The first attempt of the season, I suspect, had failed in a more secluded place under the hill; so the pair had come up nearer the house for protection. The male sang in the trees near by for several days before I chanced to see the nest. The very morning, I think, it was finished, I saw a red squirrel exploring a tree but a few yards away; he probably knew what the singing meant as well as I did. I did not see the inside of the nest, for it was almost instantly deserted, the female having probably laid a single egg, which the squirrel had devoured.

If I were a bird, in building my nest I should follow the example of the bobolink, placing it in the midst of a broad meadow, where there was no spear of grass, or flower or growth unlike another to mark its site. I judge that the bobolink escapes the dangers to which I have adverted as few or no other birds do. Unless the mowers come along at an earlier date than she has anticipated, that is, before July 1st, or a skunk goes nosing through the grass, which is unusual, she is as safe as bird well can be in the great open of nature. She selects the most monotonous and uniform place she can find amid the daisies or the timothy and clover, and places her simple structure upon the ground in the midst of it. There is no concealment, except as the great conceals the little, as the desert conceals the pebble, as the myriad conceals the unit. You may find the nest once, if your course chances to lead you across it and your eye is quick enough to note the silent brown bird as she darts quickly away; but step three paces in the wrong direction, and your search will probably be fruitless. My friend and I found a nest by accident one day, and then lost it again one minute afterward. I moved away a few yards to be sure of the mother-bird, charging my friend not to stir from his tracks. When I returned, he had moved two paces, he said (he had really moved four), and we spent a half hour stooping over the daisies and the buttercups, looking for the lost clew. We grew desperate, and fairly felt the ground all over with our hands, but without avail. I marked the spot with a bush, and came the next day, and with the bush as a centre, moved about it in slowly increasing circles, covering, I thought, nearly every inch of ground with my feet, and laying hold of it with all the visual power that I could command, till my patience was exhausted, and I gave up, baffled. I began to doubt the ability of the parent birds themselves to find it, and so secreted myself and watched. After much delay, the male bird appeared with food in his beak, and satisfying himself that the coast was clear, dropped into the grass which I had trodden down in my search. Fastening my eye upon a particular meadow-lily, I walked straight to the spot, bent down, and gazed long and intently into the grass. Finally my eye separated the nest and its young from its surroundings. My foot had barely missed them in my search, but by how much they had escaped my eye I could not tell. Probably not by distance at all, but simply by unrecognition. They were virtually invisible. The dark gray and yellowish brown dry grass and stubble of the meadow-bottom were exactly copied in the color of the half-fledged young. More than that, they hugged the nest so closely and formed such a compact mass, that though there were five of them, they preserved the unit of expression,—no single head or form was defined; they were one, and that one was without shape or color, and not separable, except by closest scrutiny, from the one of the meadow-bottom. That nest prospered, as bobolinks' nests doubtless generally do; for, notwithstanding the enormous slaughter of the birds during their fall migrations by Southern sportsmen, the bobolink appears to hold its own, and its music does not diminish in our Northern meadows.

Birds with whom the struggle for life is the sharpest seem to be more prolific than those whose nest and young are exposed to fewer dangers. The robin, the sparrow, the pewee, etc., will rear, or make the attempt to rear, two and sometimes three broods in a season; but the bobolink, the oriole, the kingbird, the goldfinch, the cedar-bird, the birds of prey, and the woodpeckers, that build in safe retreats, in the trunks of trees, have usually but a single brood. If the boblink reared two broods, our meadows would swarm with them.

I noted three nests of the cedar-bird in August in a single

orchard, all productive, but all with one or more unfruitful eggs in them. The cedar-bird is the most silent of our birds having but a single fine note, so far as I have observed, but its manners are very expressive at times. No bird known to me is capable of expressing so much silent alarm while on the nest as this bird. As you ascend the tree and draw near it, it depresses its plumage and crest, stretches up its neck, and becomes the very picture of fear. Other birds, under like circumstances, hardly change their expression at all till they launch into the air, when by their voice they express anger rather than alarm.

I have referred to the red squirrel as a destroyer of the eggs and young of birds. I think the mischief it does in this respect can hardly be over estimated. Nearly all birds look upon it as their enemy, and attack and annoy it when it appears near their breeding haunts. Thus, I have seen the pewee, the cuckoo, the robin, and the wood-thrush pursuing it with angry voice and gestures. A friend of mine saw a pair of robins attack one in the top of a tall tree so vigorously that they caused it to lose its hold, when it fell to the ground, and was so stunned by the blow as to allow him to pick it up. If you wish the birds to breed and thrive in your orchard and groves, kill every red squirrel that infests the place; kill every weasel also. The weasel is a subtle and arch enemy of the birds. It climbs trees and explores them with great ease and nimbleness. I have seen it do so on several occasions. One day my attention was arrested by the angry notes of a pair of brown-thrashers that were flitting from bush to bush along an old stone row in a remote field. Presently I saw what it was that excited them—three large red weasels, or ermines coming along the stone wall, and leisurely and half playfully exploring every tree that stood near it. They had probably robbed the thrashers. They would go up the trees with great ease, and glide serpent-like out upon the main branches. When they descended the tree they were unable to come straight down, like a squirrel, but went around it spirally. How boldly they thrust their heads out of the wall, and eyed me and sniffed me, as I drew near,—their round, thin ears, their prominent, glistening, bead-like eyes, and the curving, snake-like motions of the head and neck being very noticeable. They looked like blood-suckers and egg-suckers. They suggested something extremely remorseless and cruel. One could understand the alarm of the rats when they discover one of these fearless, subtle, and circumventing creatures threading their holes. To flee must be like trying to escape death itself. I was one day standing in the woods upon a flat stone, in what at certain seasons was the bed of a stream, when one of these weasels came undulating along and ran under the stone upon which I was standing. As I remained motionless, he thrust his wedge-shaped head, and turned it back above the stone as if half in mind to seize my foot; then he drew back, and presently went his way. These weasels often hunt in packs like the British stoat. When I was a boy, my father one day armed me with an old musket and sent me to shoot chipmunks around the corn. While watching the squirrels, a troop of weasels tried to cross a bar-way where I sat, and were so bent on doing it that I fired at them, boy-like, simply to thwart their purpose. One of the weasels was disabled by my shot, but the troop was not discouraged, and, after making several feints to cross, one of them seized the wounded one and bore it over, and the pack disappeared in the wall on the other side.

Let me conclude this chapter with two or three notes about this alert enemy of the birds and the lesser animals, the weasel.

A farmer one day heard a queer growling sound in the grass; on approaching the spot he saw two weasels contending over a mouse; each had hold of the mouse pulling in opposite directions, and were so absorbed in the struggle that the farmer cautiously put his hands down and grabbed them both by the back of the neck. He put them in a cage, and offered them bread and other food. This they refused to eat, but in a few days one of them had eaten the other up, picking his bones clean and leaving nothing but the skeleton.

The same farmer was one day in his cellar when two rats came out of a hole near him in great haste, and ran up the cellar wall and along its top till they came to a floor timber that stopped their progress, when they turned at bay, and looked excitedly back along the course they had come. In a moment a weasel, evidently in hot pursuit of them, came out of the hole, and seeing the farmer, checked his course and darted back. The rats had doubtless turned to give him fight, and would probably have been a match for him.

The weasel seems to track its game by scent. A hunter of my acquaintance was one day sitting in the woods, when he saw a red squirrel run with great speed up a tree near him, and out upon a long branch, from which he leaped to some rocks, and disappeared beneath them. In a moment a weasel came in full course upon his trail, ran up the tree, then out along the branch, from the end of which he leaped to the rocks as the squirrel did, and plunged beneath them.

Doubtless the squirrel fell a prey to him. The squirrel's best game would have been to have kept to the higher tree-tops, where he could easily have distanced the weasel. But beneath the rocks he stood a very poor chance. I have often wondered what keeps such an animal as the weasel in check, for weasels are quite rare. They never need go hungry, for rats and squirrels and mice and birds are everywhere. They probably do not fall a prey to any other animal, and very rarely to man. But the circumstances or agencies that check the increase of any species of animal are, as Darwin says, very obscure and but little known.

BEES.

AN IDYL OF THE HONEY-BEE.

There is no creature with which man has surrounded himself that seems so much like a product of civilization, so much like the result of development on special lines and in special fields, as the honey-bee. Indeed, a colony of bees, with their neatness and love of order, their division of labor, their public spiritedness, their thrift, their complex economies and their inordinate love of gain, seems as far removed from a condition of rude nature as does a walled city or a cathedral town. Our native bee, on the other hand, "the burly, dozing humble-bee," affects one more like the rude, untutored savage. He has learned nothing from experience. He lives from hand to mouth. He luxuriates in time of plenty, and he starves in times of scarcity. He lives in a rude nest or in a hole in the ground, and in small communities; he builds a few deep cells or sacks in which he stores a little honey and bee-bread for his young, but as a worker in wax he is of the most primitive and awkward. The Indian regarded the honey-bee as an ill-omen. She was the white man's fly. In fact she was the epitome of the white man himself. She has the white man's craftiness, his industry, his architectural skill, his neatness and love of system, his foresight; and above all his eager, miserly habits. The honeybee's great ambition is to be rich, to lay up great stores, to possess the sweet of every flower that blooms. She is more than provident. Enough will not satisfy her, she must have all she can get by hook or by crook. She comes from the oldest

48

country, Asia, and thrives best in the most fertile and long-settled lands.

Yet the fact remains that the honey-bee is essentially a wild creature, and never has been and cannot be thoroughly domesticated. Its proper home is the woods, and thither every new swarm counts on going; and thither many do go in spite of the care and watchfulness of the bee-keeper. If the woods in any given locality are deficient in trees with suitable cavities, the bees resort to all sorts of makeshifts; they go into chimneys, into barns and outhouses, under stones, into rocks, and so forth. Several chimneys in my locality with disused flues are taken possession of by colonies of bees nearly every season. One day, while bee-hunting, I developed a line that went toward a farm-house where I had reason to believe no bees were kept. I followed it up and questioned the farmer about his bees. He said he kept no bees, but that a swarm had taken possession of his chimney, and another had gone under the clapboards in the gable end of his house. He had taken a large lot of honey out of both places the year before. Another farmer told me that one day his family had seen a number of bees examining a knot-hole in the side of his house; the next day as they were sitting down to dinner their attention was attracted by a loud humming noise, when they discovered a swarm of bees settling upon the side of the house and pouring into the knot-hole. In subsequent years other swarms came to the same place.

Apparently, every swarm of bees before it leaves the parent hive sends out exploring parties to look up the future home. The woods and groves are searched through and through, and no doubt the privacy of many a squirrel and many a wood mouse is intruded upon. What cozy nooks and retreats they do spy out, so much more attractive than the painted hive in the garden, so much cooler in summer and so much warmer in winter!

The bee is in the main an honest citizen; she prefers legitimate to illegitimate business; she is never an outlaw until her proper sources of supply fail; she will not touch honey as long as honey-yielding flowers can be found; she always prefers to go to the fountain-head, and dislikes to take her sweets at second hand. But in the fall, after the flowers have failed, she can be tempted. The bee-hunter takes advantage of this fact; he betrays her with a little honey. He wants to steal her stores, and he first encourages her to steal his, then follows the thief home with her booty. This is the whole trick of the bee-hunter. The bees never suspect his game, else by taking a circuitous route they could easily baffle him. But the honey-bee has absolutely no wit or cunning outside of her special gifts as a gatherer and storer of honey. She is a simple-minded creature, and can be imposed upon by any novice. Yet it is not every novice that can find a bee-tree. The sportsman may track his game to its retreat by the aid of his dog, but in hunting the honey-bee one must be his own dog, and track his game through an element in which it leaves no trail. It is a task for a sharp, quick eye, and may test the resources of the best wood-craft. One autumn when I devoted much time to this pursuit, as the best means of getting at nature and the open-air exhilaration, my eye became so trained that bees were nearly as easy to it as birds. I saw and heard bees wherever I went. One day, standing on a street corner in a great city, I saw above the trucks and the traffic a line of bees carrying off sweets from some grocery or confectionery shop.

One looks upon the woods with a new interest when he suspects they hold a colony of bees. What a pleasing secret it is; a tree with a heart of comb-honey, a decayed oak or maple with a bit of Sicily or Mount Hymettus stowed away in its trunk or branches; secret chambers where lies hidden the wealth of ten thousand little freebooters, great nuggets and wedges of precious ore gathered with risk and labor from every field and wood about.

But if you would know the delights of bee-hunting, and how many sweets such a trip yields beside honey, come with me some bright, warm, late September or early October day. It is the golden season of the year, and any errand or pursuit that takes us abroad upon the hills or by the painted woods and along the amber colored streams at such a time is enough. So, with haversacks filled with grapes and peaches and apples and a bottle of milk,—for we shall not be home to dinner,—and armed with a compass, a hatchet, a pail, and a box with a piece of comb-honey neatly fitted into it—any box the size of your hand with a lid will do nearly as well as the elaborate and ingenious contrivance of the regular bee-hunter—we sally forth. Our course at first lies along the highway, under great chestnut-trees whose nuts are just dropping, then through an orchard and across a little creek, thence gently rising through a long series of cultivated fields toward some high, uplying land, behind which rises a rugged wooded ridge or mountain, the most sightly point in all this section. Behind this ridge for several miles the country is wild, wooded, and rocky, and is no doubt the home of many wild swarms of bees. What a gleeful uproar the robins, cedar-birds, high-holes, and cow black-birds make amid the black cherry-trees as we pass along. The raccoons, too, have been here after black cherries, and we see their marks at various points. Several crows are walking about a newly sowed wheat field we pass through, and we pause to note their graceful movements and glossy coats. I have seen no bird walk the ground with just the same air the crow does. It is not exactly pride; there is no strut or swagger in it, though perhaps just a little condescension; it is the contented, complaisant, and self-possessed gait of a lord over his domains. All these acres are mine, he says, and all these crops; men plow and sow for me, and I stay here or go there, and find life sweet and good wherever I am. The hawk looks awkward and out of place on the ground; the game birds hurry and skulk, but the crow is at home and treads the earth as if there were none to molest him or make him afraid.

The crows we have always with us, but it is not every day or every season that one sees an eagle. Hence I must preserve the memory of one I saw the last day I went bee-hunting. As I was laboring up the side of a mountain at the head of a valley, the noble bird sprang from the top of a dry tree above me and came sailing directly over my head. I saw him bend his eye down upon me, and I could hear the low hum of his plumage, as if the web off every quill in his great wings vibrated in his strong, level flight. I watched him as long as my eye could hold him. When he was fairly clear of the mountain he began that sweeping spiral movement in which he climbs the sky. Up and up he went without once breaking his majestic poise till he appeared to sight some far-off alien geography, when he bent his course thitherward and gradually vanished in the blue depths. The eagle is a bird of large ideas, he embraces long distances; the continent is his home. I never look upon one without emotion; I follow him with my eye as long as I can. I think of Canada, of the Great Lakes, of the Rocky Mountains, of the wild and sounding sea-coast. The waters are his, and the woods and the inaccessible cliffs. He pierces behind the veil of the storm, and his joy is height and depth and vast spaces.

We go out of our way to touch at a spring run in the edge of the woods, and are lucky to find a single scarlet lobelia lingering there. It seems almost to light up the gloom with its intense bit of color. Beside a ditch in a field beyond we find the great blue lobelia (Lobelia syphilitica), and near it amid the weeds and wild grasses and purple asters the most beautiful of our tall flowers, the fringed gentian. What a rare and delicate, almost aristocratic look the gentian has amid its coarse, unkempt surroundings. It does not lure the bee, but it lures and holds every passing human eye. If we strike through the corner of yonder woods, where the ground is moistened by hidden springs and where there is a little opening amid the trees, we shall find the closed gentian, a rare flower in this locality. I had walked this way many times before I chanced upon its retreat; and then I was following a line of bees. I lost the bees but I got the gentians. How curiously this flower looks, with its deep blue petals folded together so tightly—a bud and yet a blossom. It is the nun among our wild flowers, a form closely veiled and cloaked. The buccaneer bumble-bee sometimes tries to rifle it of its sweets. I have seen the blossom with the bee entombed in it. He had forced his way into the virgin corolla as if determined to know its secret, but he had never returned with the knowledge he had gained.

After a refreshing walk of a couple of miles we reach a point where we will make our first trial—a high stone wall that runs parallel with the wooded ridge referred to, and separated from it by a broad field. There are bees at work there on that goldenrod, and it requires but little maneuvering to sweep one into our box. Almost any other creature rudely and suddenly arrested in its career and clapped into a cage in this way would show great confusion and alarm. The bee is alarmed for a moment, but the bee has a passion stronger than its love of life or fear of death, namely, desire for honey, not simply to eat, but to carry home as booty. "Such rage of honey in their bosom beats," says Virgil. It is quick to catch the scent of honey in the box, and as quick to fall to filling itself. We now set the box down upon the wall and gently remove the cover. The bee is head and shoulders in one of the half-filled cells, and is oblivious to everything else about it. Come rack, come ruin, it will die at work. We step back a few paces, and sit down upon the ground so as to bring the box against the blue sky as a background. In two or three minutes the bee is seen rising slowly and heavily from the box. It seems loath to leave so much honey behind and it marks the place well. It mounts aloft in a rapidly increasing spiral, surveying the near and minute objects first, then the larger and more distant, till having circled about the spot five or six times and taken all its bearings it darts away for home. It is a good eye that holds fast to the bee till it is fairly off. Sometimes one's head will swim following it, and often one's eyes are put out by the sun. This bee gradually drifts down the hill, then strikes away toward a farmhouse half a mile away, where I know bees are kept. Then we try another and another, and the third bee, much to our satisfaction, goes straight toward the woods. We could see the brown speck against the darker background for many yards. The regular bee-hunter professes to be able to tell a wild bee from a tame one by the color, the former, he says, being lighter. But there is no difference; they are both alike in color and in manner. Young bees are lighter than old, and that is all there is of it. If a bee lived many years in the woods it would doubtless come to have some distinguishing marks, but the life of a bee is only a few months at the farthest, and no change is wrought in this brief time.

Our bees are all soon back, and more with them, for we have touched the box here and there with the cork of a bottle of anise oil, and this fragrant and pungent oil will attract bees half a mile or more. When no flowers can be found, this is the quickest way to obtain a bee.

It is a singular fact that when the bee first finds the hunter's box its first feeling is one of anger; it is as mad as a hornet; its tone changes, it sounds its shrill war trumpet and darts to and fro, and gives vent to its rage and indignation in no uncertain manner. It seems to scent foul play at once. It says, "Here is robbery; here is the spoil of some hive, may be my own," and its blood is up. But its ruling passion soon comes to the surface, its avarice gets the better of its indignation, and it seems to say, "Well, I had better take possession of this and carry it home." So after many feints and approaches and dartings off with a loud angry hum as if it would none of it, the bee settles down and fills itself.

It does not entirely cool off and get soberly to work till it has made two or three trips home with its booty. When other bees come, even if all from the same swarm, they quarrel and dispute over the box, and clip and dart at each other like bantam cocks. Apparently the ill feeling which the sight of the honey awakens is not one of jealousy or rivalry, but wrath.

A bee will usually make three or four trips from the hunter's box before it brings back a companion. I suspect the bee does not tell its fellows what it has found, but that they smell out the secret; it doubtless bears some evidence with it upon its feet or proboscis that it has been upon honey-comb and not upon flowers, and its companions take the hint and follow, arriving always many seconds behind. Then the quantity and quality of the booty would also betray it. No doubt, also, there are plenty of gossips about a hive that note and tell everything. "Oh, did you see that? Peggy Mel came in a few moments ago in great haste, and one of the up-stairs packers says she was loaded till she groaned with apple-blossom honey which she deposited, and then rushed off again like mad. Apple-blossom honey in October! Fee, fi, fo, fum! I smell something! Let's after."

In about half an hour we have three well-defined lines of bees established—two to farm-houses and one to the woods, and our box is being rapidly depleted of its honey. About every fourth bee goes to the woods, and now that they have learned the way thoroughly they do not make the long preliminary whirl above the box, but start directly from it. The woods are rough and dense and the hill steep, and we do not like to follow the line of bees until we have tried at least to settle the problem as to the distance they go into the woods-whether the tree is on this side of the ridge or in the depth of the forest on the other side. So we shut up the box when it is full of bees and carry it about three hundred yards along the wall from which we are operating. When liberated, the bees, as they always will in such cases, go off in the same directions they have been going; they do not seem to know that they have been moved. But other bees have followed our scent, and it is not many minutes before a second line to the woods is established. This is called cross-lining the bees. The new line makes a sharp angle with the other line, and we know at once that the tree is only a few rods into the woods. The two lines we have established form two sides of a triangle of which the wall is the base; at the apex of the triangle, or where the two lines meet in the woods, we are sure to find the tree. We quickly follow up these lines, and where they cross each other on the side of the hill we scan every tree closely. I pause at the foot of an oak and examine a hole near the root; now the bees are in this tree and their entrance is on the upper side near the ground, not two feet from the hole I peer into, and yet so quiet and secret is their going and coming that I fail to discover them and pass on up the hill. Failing in this direction, I return to the oak again, and then perceive the bees going out in a small crack in the tree. The bees do not know they are found out and that the game is in our hands, and are as oblivious of our presence as if we were ants or crickets. The indications are that the swarm is a small one, and the store of honey trifling. In "taking up" a bee-tree it is usual first to kill or stupefy the bees with the fumes of burning sulfur or with tobacco smoke. But this course is impracticable on the present occasion, so we boldly and ruthlessly assault the tree with an ax we have procured. At the first blow the bees set up a loud buzzing, but we have no mercy, and the side of the cavity is soon cut away and the interior with its white-yellow mass of comb-honey is exposed, and not a bee strikes a blow in defense of its all. This may seem singular, but it has nearly always been my experience. When a swarm of bees are thus rudely assaulted with an ax, they evidently think the end of the world has come, and, like true misers as they are, each one seizes as much of the treasure as it can hold; in other words they all fall to and gorge themselves with honey, and calmly await the issue. When in this condition they make no defense and will not sting unless taken hold of. In fact they are as harmless as flies. Bees are always to be managed with boldness and decision.

Any half-way measures, any timid poking about, any feeble attempts to reach their honey, are sure to be quickly resented. The popular notion that bees have a special antipathy toward certain persons and a liking for certain others has only this fact at the bottom of it; they will sting a person who is afraid of them and goes skulking and dodging about, and they will not sting a person who faces them boldly and has no dread of them. They are like dogs. The way to disarm a vicious dog is to show him you do not fear him; it is his turn to be afraid then. I never had any dread of bees and am seldom stung by them. I have climbed up into a large chestnut that contained a swarm in one of its cavities and chopped them out with an ax, being obliged at times to pause and brush the bewildered bees from my hands and face, and not been stung once. I have chopped a swarm out of an apple-tree in June and taken out the cards of honey and arranged them in a hive, and then dipped out the bees with a dipper, and taken the whole home with me in pretty good condition, with scarcely any opposition on the part of the bees. In reaching your hand into the cavity to detach and remove the comb you are pretty sure to get stung, for when you touch the "business end" of a bee, it will sting even though its head be off. But the bee carries the antidote to its own poison. The best remedy for bee sting is honey, and when your hands are besmeared with honey, as they are sure to be on such occasions, the wound is scarcely more painful than the prick of a pin. Assault your bee-tree, then, boldly with your ax, and you will find that when the honey is exposed every bee has surrendered and the whole swarm is cowering in helpless bewilderment and terror. Our tree yields only a few pounds of honey, not enough to have lasted the swarm till January, but no matter; we have the less burden to carry.

In the afternoon we go nearly half a mile farther along the ridge to a cornfield that lies immediately in front of the highest point of the mountain. The view is superb; the ripe autumn landscape rolls away to the east, cut through by the great placid river; in the extreme north the wall of the Catskills stands out clear and strong, while in the south the mountains of the Highlands bound the view. The day is warm and the bees are very busy there in that neglected corner of the field, rich in asters, flea-bane, and golden-rod. The corn has been cut, and upon a stout, but a few rods from the woods, which here drop quickly down from the precipitous heights, we set up our bee-box, touched again with the pungent oil. In a few moments a bee has found it; she comes up to leeward, following the scent. On leaving the box she goes straight toward the woods. More bees quickly come, and it is not long before the line is well established. Now we have recourse to the same tactics we employed before, and move along the ridge to another field to get our cross line. But the bees still go in almost the same direction they did from the corn stout. The tree is then either on the top of the mountain or on the other or west side of it. We hesitate to make the plunge into the woods and seek to scale those precipices, for the eye can plainly see what is before us. As the afternoon sun gets lower the bees are seen with wonderful distinctness. They fly toward and under the sun and are in a strong light, while the near woods which form the background are in deep shadow. They look like large luminous motes. Their swiftly vibrating, transparent wings surround their bodies with a shining nimbus that makes them visible for a long distance. They seem magnified many times. We see them bridge the little gulf between us and the woods, then rise up over the tree-tops with their burdens, swerving neither to the right hand nor to the left. It is almost pathetic to see them labor so, climbing the mountain and unwittingly guiding us to their treasures. When the sun gets down so that his direction corresponds exactly with the course of the bees, we make the plunge. It proves even harder climbing than we had anticipated; the mountain is faced by a broken and irregular wall of rock, up which we pull ourselves slowly and cautiously by main strength. In half an hour, the perspiration streaming from every pore, we reach the summit. The trees here are all small, a second growth, and we are soon convinced the bees are not here. Then down we go on the other side, clambering down the rocky stairways till we reach quite a broad plateau that forms something like the shoulder of the mountain. On the brink of this there are many large hemlocks, and we scan them closely and rap upon them with our ax. But not a bee is seen or heard; we do not seem as near the tree as we were in the fields below; yet if some divinity would only whisper the fact to us we are within a few rods of the coveted prize, which is not in one of the large hemlocks or oaks that absorb our attention, but in an old stub or stump not six feet high, and which we have seen and passed several times without giving it a thought. We go farther down the mountain and beat about to the right and left and get entangled in brush and arrested by precipices, and finally as the day is nearly spent, give up the search and leave the woods quite baffled, but resolved to return on the morrow. The next day we come back and commence operations in an opening in the woods well down on the side of the mountain, where we gave up the search. Our box is soon swarming with the eager bees, and they go back toward the summit we have passed. We follow back and establish a new line where the ground will permit; then another and another, and yet the riddle is not solved. One time we are south of them, then north, then the bees get up through the trees and we cannot tell where they go. But after much searching, and after the mystery seems rather to deepen than to clear up, we chance to pause beside the old stump. A bee comes out of a small opening, like that made by ants in decayed wood, rubs its eyes and examines its antennae as bees always do before leaving their hive, then takes flight. At the same instant several bees come by us loaded with our honey and settle home with that peculiar low complacent buzz of the well-filled insect. Here then is our idyl, our bit of Virgil and Theocritus, in a decayed stump of a hemlock tree. We could tear it open with our hands, and a bear would find it an easy prize, and a rich one too, for we take from it fifty pounds of excellent honey. The bees have been here many years, and have of course sent out swarm after swarm into the wilds. They have protected themselves against the weather and strengthened their shaky habitation by a copious use of wax.

When a bee-tree is thus "taken up" in the middle of the day, of course a good many bees are away from home and have not heard the news. When they return and find the ground flowing with honey, and piles of bleeding combs lying about, they apparently do not recognize the place, and their first instinct is to fall to and fill themselves; this done, their next thought is to carry it home, so they rise up slowly through the branches of the trees till they have attained an altitude that enables them to survey the scene, when they seem to say, "Why, this is home," and down they come again; beholding the wreck and ruins once more they still think there is some mistake, and get up a second or a third time and then drop back pitifully as before. It is the most pathetic sight of all, the surviving and bewildered bees struggling to save a few drops of their wasted treasures.

Presently, if there is another swarm in the woods, robber-bees appear. You may know them by their saucy, chiding, devil-may-care hum. It is an ill wind that blows nobody good, and they make the most of the misfortune of their neighbors; and thereby pave the way for their own ruin. The hunter marks their course and the next day looks them up. On this occasion the day was hot and the honey very fragrant, and a line of bees was soon established S. S. W. Though there was much refuse honey in the old stub, and though little golden rills trickled down the hill from it, and the near branches and saplings were besmeared with it where we wiped our murderous hands, yet not a drop was wasted. It was a feast to which not only honey-bees came, but bumble-bees, wasps, hornets, flies, ants. The bumble-bees, which at this season are hungry vagrants with no fixed place of abode, would gorge themselves, then creep beneath the bits of empty comb or fragments of bark and pass the night, and renew the feast next day. The bumble-bee is an insect of which the bee-hunter sees much. There are all sorts and sizes of them. They are dull and clumsy compared with the honey-bee. Attracted in the fields by the bee-hunter's box, they will come up the wind on the scent and blunder into it in the most stupid, lubberly fashion.

The honey-bee that licked up our leavings on the old stub belonged to a swarm, as it proved, about half a mile farther down the ridge, and a few days afterward fate overtook them, and their stores in turn became the prey of another swarm in the vicinity, which also tempted Providence and were overwhelmed. The first mentioned swarm I had lined from several points, and was following up the clew over rocks and through gulleys, when I came to where a large hemlock had been felled a few years before and a swarm taken from a cavity near the top of it; fragments of the old comb were yet to be seen. A few yards away stood another short, squatty hemlock, and I said my bees ought to be there. As I paused near it I noticed where the tree had been wounded with an ax a couple of feet from the ground many years before. The wound had partially grown over, but there was an opening there that I did not see at the first glance. I was about to pass on when a bee passed me making that peculiar shrill, discordant hum that a bee makes when besmeared with honey. I saw it alight in the partially closed wound and crawl home; then came others and others, little bands and squads of them heavily freighted with honey from the box. The tree was about twenty inches through and hollow at the butt, or from the ax mark down. This space the bees had completely filled with honey. With an ax we cut away the outer ring of live wood and exposed the treasure. Despite the utmost care, we wounded the comb so that little rills of the golden liquid issued from the root of the tree and trickled down the hill.

The other bee-tree in the vicinity, to which I have referred, we found one warm November day in less than half an hour after entering the woods. It also was a hemlock, that stood in a niche in a wall of hoary, moss-covered rocks thirty feet high. The tree hardly reached to the top of the precipice. The bees entered a small hole at the root, which was seven or eight feet from the ground. The position was a striking one. Never did apiary have a finer outlook or more rugged surroundings. A black, wood-embraced lake lay at our feet; the long panorama of the Catskills filled the far distance, and the more broken outlines of the Shawangunk range filled the rear. On every hand were precipices and a wild confusion of rocks and trees.

The cavity occupied by the bees was about three feet and a half long and eight or ten inches in diameter. With an ax we cut away one side of the tree and laid bare its curiously wrought heart of honey. It was a most pleasing sight. What winding and devious ways the bees had through their palace! What great masses and blocks of snow-white comb there were! Where it was sealed up, presenting that slightly dented, uneven surface, it looked like some precious ore. When we carried a large pail full of it out of the woods, it seemed still more like ore.

Your native bee-hunter predicates the distance of the tree by the time the bee occupies in making its first trip. But this is no certain guide. You are always safe in calculating that the tree is inside of a mile, and you need not as a rule look for your bee's return under ten minutes. One day I picked up a bee in an opening in the woods and gave it honey, and it made three trips to my box with an interval of about twelve minutes between them; it returned alone each time; the tree, which I afterward found, was about half a mile distant.

In lining bees through the woods, the tactics of the hunter are to pause every twenty or thirty rods, lop away the branches or cut down the trees, and set the bees to work again. If they still go forward, he goes forward also and repeats his observations till the tree is found or till the bees turn and come back upon the trail. Then he knows he has passed the tree, and he retraces his steps to a convenient distance and tries again, and thus quickly reduces the space to be looked over till the swarm is traced home. On one occasion, in a wild rocky wood, where the surface alternated between deep gulfs and chasms filled with thick, heavy growths of timber and sharp, precipitous, rocky ridges like a tempest tossed sea, I carried my bees directly under their tree, and set them to work from a high, exposed ledge of rocks not thirty feet distant. One would have expected them under such circumstances to have gone straight home, as there were but few branches intervening, but they did not; they labored up through the trees and attained an altitude above the woods as if they had miles to travel, and thus baffled me for hours. Bees will always do this. They are acquainted with the woods only from the top side, and from the air above they recognize home only by land-marks here, and in every instance they rise aloft to take their bearings. Think how familiar to them the topography of the forest summits must be-an umbrageous sea or plain where every mask and point is known.

Another curious fact is that generally you will get track of a bee-tree sooner when you are half a mile from it than when you are only a few yards. Bees, like us human insects, have little faith in the near at hand; they expect to make their fortune in a distant field, they are lured by the remote and the difficult, and hence overlook the flower and the sweet at their very door. On several occasions I have unwittingly set my box within a few paces of a bee-tree and waited long for bees without getting them, when, on removing to a distant field or opening in the woods I have got a clew at once.

I have a theory that when bees leave the hive, unless there is some special attraction in some other direction, they generally go against the wind. They would thus have the wind with them when they returned home heavily laden, and with these little navigators the difference is an important one. With a full cargo, a stiff head-wind is a great hindrance, but fresh and empty-handed they can face it with more ease. Virgil says bees bear gravel stones as ballast, but their only ballast is their honey bag. Hence, when I go bee-hunting, I prefer to get to windward of the woods in which the swarm is supposed to have taken refuge.

Bees, like the milkman, like to be near a spring. They do water their honey, especially in a dry time. The liquid is then of course thicker and sweeter, and will bear diluting. Hence, old bee-hunters look for bee-trees along creeks and near spring runs in the woods. I once found a tree a long distance from any water, and the honey had a peculiar bitter flavor imparted to it, I was convinced, by rainwater sucked from the decayed and spongy hemlock tree, in which the swarm was found. In cutting into the tree, the north side of it was found to be saturated with water like a spring, which ran out in big drops, and had a bitter flavor. The bees had thus found a spring or a cistern in their own house.

Bees are exposed to many hardships and many dangers. Winds and storms prove as disastrous to them as to other navigators. Black spiders lie in wait for them as do brigands for travelers. One day as I was looking for a bee amid some golden-rod, I spied one partly concealed under a leaf. Its baskets were full of pollen, and it did not move. On lifting up the leaf I discovered that a hairy spider was ambushed there and had the bee by the throat. The vampire was evidently afraid of the bee's sting, and was holding it by the throat till quite sure of its death. Virgil speaks of the painted lizard, perhaps a species of salamander, as an enemy of the honey-bee. We have no lizard that destroys the bee; but our tree-toad, ambushed among the apple and cherry blossoms, snaps them up wholesale. Quick as lightning that subtle but clammy tongue darts forth, and the unsuspecting bee is gone. Virgil also accuses the titmouse and the woodpecker of preying upon the bees, and our kingbird has been charged with the like crime, but the latter devours only the drones. The workers are either too small and quick for it, or else it dreads their sting.

Virgil, by the way, had little more than a child's knowledge of the honey-bee. There is little fact and much fable in his fourth Georgic. If he had ever kept bees himself, or even visited an apiary, it is hard to see how he could have believed that the bee in its flight abroad carried a gravel stone for ballast:—

> *"And as when empty barks on billows float,*
> *With Sandy ballast sailors trim the boat;*
> *So bees bear gravel stones, whose poising weight*
> *Steers through the whistling winds their steady flight;"*

or that when two colonies made war upon each other they issued forth from their hives led by their kings and fought in the air, strewing the ground with the dead and dying:—

> *"Hard hailstones lie not thicker on the plain,*
> *Nor shaken oaks such show'rs of acorns rain."*

It is quite certain he had never been bee-hunting. If he had, we should have had a fifth Georgic. Yet he seems to have known that bees sometimes escaped to the woods:—

"Nor bees are lodged in hives alone, but found
In chambers of their own beneath the ground:
Their vaulted roofs are hung in pumices,
And in the rotten trunks of hollow trees."

Wild honey is as near like tame as wild bees are like their brothers in hive. The only difference is that wild honey is flavored with your adventure, which makes it a little more delectable than the domestic article.

THE PASTORAL BEES

The honey-bee goes forth from the hive in spring like the dove from Noah's ark, and it is not till after many days that she brings back the olive leaf, which in this case is a pellet of golden pollen upon each hip, usually obtained from the alder or the swamp willow. In a country where maple sugar is made, the bees get their first taste of sweet from the sap as it flows from the spiles, or as it dries and is condensed upon the sides of the buckets. They will sometimes, in their eagerness, come about the boiling place and be overwhelmed by the steam and the smoke. But bees appear to be more eager for bread in the spring than for honey; their supply of this article, perhaps, does not keep as well as their stores of the latter, hence fresh bread, in the shape of new pollen, is diligently sought for. My bees get their first supplies from the catkins of the willows. How quickly they find them out. If but one catkin opens anywhere within range, a bee is on hand that very hour to rifle it, and it is a most pleasing experience to stand near the hive some mild April day and see them come pouring in with their little baskets packed with this first fruitage of the spring. They will have new bread now; they have been to mill in good earnest; see their dusty coats, and the golden grist they bring home with them.

When a bee brings pollen into the hive, he advances to the cell in which it is to be deposited and kicks it off as one might his overalls or rubber boots, making one foot help the other; then he walks off without ever looking behind him; another bee, one of the indoor hands, comes along and rams it down with his head and packs it into the cell as the dairymaid packs butter into a firkin.

The first spring wild-flowers, whose shy faces among the dry leaves and rocks are so welcome, yield no honey. The anemone, the hepatica, the bloodroot, the arbutus, the numerous violets, the spring beauty, the corydalis, etc., woo lovers of nature, but do not woo the honey-loving bee. It requires more sun and warmth to develop the saccharine element, and the beauty of these pale striplings of the woods and groves is their sole and sufficient excuse for being. The arbutus, lying low and keeping green all winter, attains to perfume, but not to honey.

The first honey is perhaps obtained from the flowers of the red maple and the golden willow. The latter sends forth a wild, delicious perfume. The sugar maple blooms a little later, and from its silken tassels a rich nectar is gathered. My bees will not label these different varieties for me as I really wish they would. Honey from the maples, a tree so clean and wholesome, and full of such virtues every way, would be something to put one's tongue to. Or that from the blossoms of the apple, the peach, the cherry, the quince, the currant,—one would like a card of each of these varieties to note their peculiar qualities. The apple-blossom is very important to the bees. A single swarm has been known to gain twenty pounds in weight during its continuance. Bees love the ripened fruit, too, and in August and September will suck themselves tipsy upon varieties such as the sops-of-wine.

The interval between the blooming of the fruit-trees and that of the clover and the raspberry is bridged over in many localities by the honey locust. What a delightful summer murmur these trees send forth at this season. I know nothing about the quality of the honey, but it ought to keep well. But when the red raspberry blooms, the fountains of plenty are unsealed indeed; what a commotion about the hives then, especially in localities where it is extensively cultivated, as in places along the Hudson. The delicate white clover, which begins to bloom about the same time, is neglected; even honey itself is passed by for this modest colorless, all but odorless flower. A field of these berries in June sends forth a continuous murmur like that of an enormous hive. The honey is not so white as that obtained from clover but it is easier gathered; it is in shallow cups while that of the clover is in deep tubes. The bees are up and at it before sunrise, and it takes a brisk shower to drive them in. But the clover blooms later and blooms everywhere, and is the staple source of supply of the finest quality of honey. The red clover yields up its stores only to the longer proboscis of the bumble-bee, else the bee pasturage of our agricultural districts would be unequaled. I do not know from what the famous honey of Chamouni in the Alps is made, but it can hardly surpass our best products. The snow-white honey of Anatolia in Asiatic Turkey, which is regularly sent to Constantinople for the use of the grand seignior and the ladies of his seraglio, is obtained from the cotton plant, which makes me think that the white clover does not flourish these. The white clover is indigenous with us; its seeds seem latent in the ground, and the application of certain stimulants to the soil, such as wood ashes, causes them to germinate and spring up.

The rose, with all its beauty and perfume, yields no honey to the bee, unless the wild species be sought by the bumble-bee.

Among the humbler plants, let me not forget the dandelion that so early dots the sunny slopes, and upon which the bee languidly grazes, wallowing to his knees in the golden but not over-succulent pasturage. From the blooming rye and wheat the bee gathers pollen, also from the obscure blossoms of Indian corn. Among weeds, catnip is the great favorite. It lasts nearly the whole season and yields richly. It could no doubt be profitably cultivated in some localities, and catnip honey would be a novelty in the market. It would probably partake of the aromatic properties of the plant from which it was derived.

Among your stores of honey gathered before midsummer, you may chance upon a card, or mayhap only a square inch or two of comb, in which the liquid is as transparent as water, of a delicious quality, with a slight flavor of mint. This is the product of the linden or basswood, of all the trees in our forest the one most beloved by the bees. Melissa, the goddess of honey, has placed her seal upon this tree. The wild swarms in the woods frequently reap a choice harvest from it. I have seen a mountain side thickly studded with it, its straight, tall, smooth, light-gray shaft carrying its deep-green crown far aloft, like the tulip-tree or the maple.

In some of the Northwestern States there are large forests of it, and the amount of honey reported stored by strong swarms in this section during the time the tree is in bloom is quite incredible. As a shade and ornamental tree the linden is fully equal to the maple, and if it were as extensively planted and cared for, our supplies of virgin honey would be greatly increased. The famous honey of Lithuania in Russia is the product of the linden.

It is a homely old stanza current among bee folk that—

> *"A swarm of bees in May*
> *Is worth a load of hay;*
> *A swarm of bees in June*
> *Is worth a silver spoon;*
> *But a swarm in July*
> *Is not worth a fly."*

A swarm in May is indeed a treasure; it is, like an April baby, sure to thrive, and will very likely itself send out a swarm a month or two later; but a swarm in July is not to be despised; it will store no clover or linden honey for the "grand seignior and the ladies of his seraglio," but plenty of the rank and wholesome poor

man's nectar, the sun-tanned product of the plebeian buckwheat. Buckwheat honey is the black sheep in this white flock, but there is spirit and character in it. It lays hold of the taste in no equivocal manner, especially when at a winter breakfast it meets its fellow, the russet buckwheat cake. Bread with honey to cover it from the same stalk is double good fortune. It is not black, either, but nut-brown, and belongs to the same class of goods as Herrick's
  "Nut-brown mirth and russet wit."

How the bees love it, and they bring the delicious odor of the blooming plant to the hive with them, so that in the moist warm twilight the apiary is redolent with the perfume of buckwheat.

Yet evidently it is not the perfume of any flower that attracts the bees; they pay no attention to the sweet-scented lilac, or to heliotrope, but work upon sumach, silkweed, and the hateful snapdragon. In September they are hard pressed, and do well if they pick up enough sweet to pay the running expenses of their establishment. The purple asters and the golden-rod are about all that remain to them.

Bees will go three or four miles in quest of honey, but it is a great advantage to move the hive near the good pasturage, as has been the custom from the earliest times in the Old World. Some enterprising person, taking a hint perhaps from the ancient Egyptians, who had floating apiaries on the Nile, has tried the experiment of floating several hundred colonies north on the Mississippi, starting from New Orleans and following the opening season up, thus realizing a sort of perpetual May or June, the chief attraction being the blossoms of the river willow, which yield honey of rare excellence. Some of the bees were no doubt left behind, but the amount of virgin honey secured must have been very great. In September they should have begun the return trip, following the retreating summer South.

It is the making of the wax that costs with the bee. As with the poet, the form, the receptacle, gives him more trouble than the sweet that fills it, though, to be sure, there is always more or less empty comb in both cases. The honey he can have for the gathering, but the wax he must make himself—must evolve from his own inner consciousness. When wax is to be made the wax-makers fill themselves with honey and retire into their chamber for private meditation; it is like some solemn religious rite; they take hold of hands, or hook themselves together in long lines that hang in festoons from the top of the hive, and wait for the miracle to transpire. After about twenty-four hours their patience is rewarded, the honey is turned into wax, minute scales of which are secreted from between the rings of the abdomen of each bee; this is taken off and from it the comb is built up. It is calculated that about twenty-five pounds of honey are used in elaborating one pound of comb, to say nothing of the time that is lost. Hence the importance in an economical point of view, of a recent device by which the honey is extracted and the comb returned intact to the bees. But honey without the comb is the perfume without the rose,—it is sweet merely, and soon degenerates into candy. Half the delectableness is in breaking down these frail and exquisite walls yourself, and tasting the nectar before it has lost its freshness by the contact with the air. Then the comb is a sort of shield or foil that prevents the tongue from being overwhelmed by the shock of the sweet.

The drones have the least enviable time of it. Their foothold in the hive is very precarious. They look like the giants, the lords of the swarm, but they are really the tools. Their loud, threatening hum has no sting to back it up, and their size and noise make them only the more conspicuous marks for the birds.

Toward the close of the season, say in July or August, the fiat goes forth that the drones must die; there is no further use for them. Then the poor creatures, how they are huddled and hustled about, trying to hide in corners and by-ways. There is no loud, defiant humming now, but abject fear seizes them. They cower like hunted criminals. I have seen a dozen or more of them wedge themselves into a small space between the glass and the comb, where the bees could not get hold of them or where they seemed to be overlooked in the general slaughter. They will also crawl outside and hide under the edges of the hive. But sooner or later they are all killed or kicked out. The drone makes no resistance, except to pull back and try to get away; but (putting yourself in his place) with one bee a-hold of your collar or the hair of your head, and another a-hold of each arm or leg, and still another feeling for your waistbands with his sting, the odds are greatly against you.

It is a singular fact, also, that the queen is made, not born. If the entire population of Spain or Great Britain were the offspring of one mother, it might be found necessary to hit upon some device by which a royal baby could be manufactured out of an ordinary one, or else give up the fashion of royalty. All the bees in the hive have a common parentage, and the queen and the worker are the same in the egg and in the chick; the patent of royalty is in the cell and in the food; the cell being much larger, and the food a peculiar stimulating kind of jelly. In certain contingencies, such as the loss of the queen with no eggs in the royal cells, the workers take the larva of an ordinary bee, enlarge the cell by taking in the two adjoining ones, and nurse it and stuff it and coddle it, till at the end of sixteen days it comes out a queen. But ordinarily, in the natural course of events, the young queen is kept a prisoner in her cell till the old queen has left with the swarm. Later on, the unhatched queen is guarded against the reigning queen, who only wants an opportunity to murder every royal scion in the hive. At this time both the queens, the one a prisoner and the other at large, pipe defiance at each other, a shrill, fine, trumpet-like note that any ear will at once recognize. This challenge, not being allowed to be accepted by either party, is followed, in a day or two by the abdication of the reigning queen; she leads out the swarm, and her successor is liberated by her keepers, who, in her time, abdicates in favor of the next younger. When the bees have decided that no more swarms can issue, the reigning queen is allowed to use her stiletto upon her unhatched sisters. Cases have been known where two queens issued at the same time, when a mortal combat ensued, encouraged by the workers, who formed a ring about them, but showed no preference, and recognized the victor as the lawful sovereign. For these and many other curious facts we are indebted to the blind Huber.

It is worthy of note that the position of the queen cells is always vertical, while that of the drones and workers is horizontal; majesty stands on its head, which fact may be a part of the secret.

The notion has always very generally prevailed that the queen of the bees is an absolute ruler, and issues her royal orders to willing subjects. Hence Napoleon the First sprinkled the symbolic bees over the imperial mantle that bore the arms of his dynasty; and in the country of the Pharaohs the bee was used as the emblem of a people sweetly submissive to the orders of its king. But the fact is, a swarm of bees is an absolute democracy, and kings and despots can find no warrant in their example. The power and authority are entirely vested in the great mass, the workers. They furnish all the brains and foresight of the colony, and administer its affairs. Their word is law, and both king and queen must obey. They regulate the swarming, and give the signal for the swarm to issue from the hive; they select and make ready the tree in the woods and conduct the queen to it.

The peculiar office and sacredness of the queen consists in the fact that she is the mother of the swarm, and the bees love and cherish her as a mother and not as a sovereign. She is the sole female bee in the hive, and the swarm clings to her because she is their life. Deprived of their queen, and of all brood from which to rear one, the swarm loses all heart and soon dies, though there be an abundance of honey in the hive.

The common bees will never use their sting upon the queen; if she is to be disposed of they starve her to death; and the queen herself will sting nothing but royalty—nothing but a rival queen.

The queen, I say, is the mother bee; it is undoubtedly complimenting her to call her a queen and invest her with regal authority, yet she is a superb creature, and looks every inch a queen. It is an event to distinguish her amid the mass of bees when the swarm alights; it awakens a thrill. Before you have seen a queen you wonder if this or that bee, which seems a little larger than its fellows, is not she, but when you once really set eyes upon her you do not doubt for a moment. You know that is the queen. That long, elegant, shining, feminine-looking creature can be none less than royalty. How beautifully her body tapers, how distinguished she looks, how deliberate her movements! The bees do not fall down before her, but caress her and touch her person. The drones or males, are large bees too, but coarse, blunt, broad-shouldered, masculine-looking. There is but one fact or incident in the life of the queen that looks imperial and authoritative: Huber relates that when the old queen is restrained in her movements by the workers, and prevented from destroying the young queens in their cells, she assumes a peculiar attitude and utters a note that strikes every bee motionless, and makes every head bow; while this sound lasts not a bee stirs, but all look abashed and humbled, yet whether the emotion is one of fear, or reverence, or of sympathy with the distress of the queen mother, is hard to determine. The moment it ceases and she advances again toward the royal cells, the bees bite and pull and insult her as before.

I always feel that I have missed some good fortune if I am away from home when my bees swarm. What a delightful summer sound it is; how they come pouring out of the hive, twenty or thirty thousand bees each striving to get out first; it is as when the dam gives way and lets the waters loose; it is a flood of bees which breaks upward into the air, and becomes a maze of whirling black lines to the eye and a soft chorus of myriad musical sounds to the ear. This way and that way they drift, now contracting, now expanding, rising, sinking, growing thick about some branch or bush, then dispersing and massing at some other point, till finally they begin to alight in earnest, when in a few moments the whole swarm is collected upon the branch, forming a bunch perhaps as large as a two-gallon measure. Here they will hang from one to three or four hours, or until a suitable tree in the woods is looked up, when, if they have not been offered a hive in the mean time, they are up and off. In hiving them, if any accident happens to the queen the enterprise miscarries at once. One day I shook a swarm from a small pear-tree into a tin pan, set the pan down on a shawl spread beneath the tree, and put the hive over it. The bees presently all crawled up into it, and all seemed to go well for ten or fifteen minutes, when I observed that something was wrong; the bees began to buzz excitedly and to rush about in a bewildered manner, then they took to the wing and all returned to the parent stock. On lifting up the pan, I found beneath it the queen with three or four other bees. She had been one of the first to fall, had missed the pan in her descent, and I had set it upon her. I conveyed her tenderly back to the hive, but either the accident terminated fatally with her or else the young queen had been liberated in the interim, and one of them had fallen in combat, for it was ten days before the swarm issued a second time.

No one, to my knowledge, has ever seen the bees house-hunting in the woods. Yet there can be no doubt that they look up new quarters either before or on the day the swarm issues. For all bees are wild bees and incapable of domestication; that is, the instinct to go back to nature and take up again their wild abodes in the trees is never eradicated. Years upon years of life in the apiary seems to have no appreciable effect towards their final, permanent domestication. That every new swarm contemplates migrating to the woods, seems confirmed by the fact that they will only come out when the weather is favorable to such an enterprise, and that a passing cloud or a sudden wind, after the bees are in the air, will usually drive them back into the parent hive. Or an attack upon them with sand or gravel, or loose earth or water, will quickly cause them to change their plans. I would not even say but that, when the bees are going off, the apparently absurd practice, now entirely discredited by regular bee-keepers but still resorted to by unscientific folk, of beating upon tin pans, blowing horns, and creating an uproar generally, might not be without good results. Certainly not by drowning the "orders" of the queen, but by impressing the bees as with some unusual commotion in nature. Bees are easily alarmed and disconcerted, and I have known runaway swarms to be brought down by a farmer ploughing in the field who showered them with handfuls of loose soil.

I love to see a swarm go off—if it is not mine, and if mine must go I want to be on hand to see the fun. It is a return to first principles again by a very direct route. The past season I witnessed two such escapes. One swarm had come out the day before, and, without alighting, had returned to the parent hive—some hitch in the plan, perhaps, or may be the queen had found her wings too weak. The next day they came out again, and were hived. But something offended them, or else the tree in the woods—perhaps some royal old maple or birch holding its head high above all others, with snug, spacious, irregular chambers and galleries—had too many attractions; for they were presently discovered filling the air over the garden, and whirling excitedly around. Gradually they began to drift over the street; a moment more, and they had become separated from the other bees, and, drawing together in a more compact mass or cloud, away they went, a humming, flying vortex of bees, the queen in the centre, and the swarm revolving around her as a pivot,—over meadows, across creeks and swamps, straight for the heart of the mountain, about a mile distant,—slow at first, so that the youth who gave chase kept up with them, but increasing their speed till only a fox hound could have kept them in sight. I saw their pursuer laboring up the side of the mountain; saw his white shirt-sleeves gleam as he entered the woods; but he returned a few hours afterward without any clew as to the particular tree in which they had taken refuge out of the ten thousand that covered the side of the mountain.

The other swarm came out about one o'clock of a hot July day, and at once showed symptoms that alarmed the keeper, who, however, threw neither dirt nor water. The house was situated on a steep side-hill. Behind it the ground rose, for a hundred rods or so, at an angle of nearly forty-five degrees, and the prospect of having to chase them up this hill, if chase them we should, promised a good trial of wind at least; for it soon became evident that their course lay in this direction. Determined to have a hand, or rather a foot, in the chase, I threw off my coat and hurried on, before the swarm was yet fairly organized and under way. The route soon led me into a field of standing rye, every spear of which held its head above my own. Plunging recklessly forward, my course marked to those watching from below by the agitated and wriggling grain, I emerged from the miniature forest just in time to see the runaways disappearing over the top of the hill, some fifty rods in advance of me. Lining them as well as I could, I soon reached the hill-top, my breath utterly gone and the perspiration streaming from every pore of my skin. On the other side the country opened deep and wide. A large valley swept around to the north, heavily wooded at its head and on its sides. It became evident at once that the bees had made good their escape, and that whether they had stopped on one side of the valley or the other, or had indeed cleared the opposite mountain and gone into some unknown forest beyond, was entirely problematical. I turned back, therefore, thinking of the honey-laden tree that some of these forests would hold before the falling of the leaf.

I heard of a youth in the neighborhood, more lucky than myself on a like occasion. It seems that he had got well in advance of the swarm, whose route lay over a hill, as in my case, and as he neared the summit, hat in hand, the bees had just come up and were all about him. Presently he noticed them hovering about his straw hat, and alighting on his arm; and in almost as brief a time as it takes to relate it, the whole swarm had followed the queen into his hat. Being near a stone wall, he coolly deposited his prize upon it, quickly disengaged himself from the accommodating bees, and returned for a hive. The explanation of this singular circumstance no doubt is, that the queen, unused to such long and heavy flights, was obliged to alight from very exhaustion. It is not very unusual for swarms to be thus found in remote fields, collected upon a bush or branch of a tree.

When a swarm migrates to the woods in this manner, the individual bees, as I have intimated, do not move in right lines or straight forward, like a flock of birds, but round and round, like chaff in a whirlwind. Unitedly they form a humming, revolving, nebulous mass, ten or fifteen feet across, which keeps just high enough to clear all obstacles, except in crossing deep valleys, when, of course, it may be very high. The swarm seems to be guided by a line of couriers, which may be seen (at least at the outset) constantly going and coming. As they take a direct course, there is always some chance of following them to the tree, unless they go a long distance, and some obstruction, like a wood, or a swamp, or a high hill, intervenes—enough chance, at any rate, to stimulate the lookers-on to give vigorous chase as long as their wind holds out. If the bees are successfully followed to their retreat, two plans are feasible: either to fell the tree at once, and seek to hive them, perhaps bring them home in the section of the tree that contains the cavity; or to leave the tree till fall, then invite your neighbors, and go and cut it, and see the ground flow with honey. The former course is more business-like; but the latter is the one usually recommended by one's friends and neighbors.

Perhaps nearly one third of all the runaway swarms leave when no one is about, and hence are unseen and unheard, save, perchance, by some distant laborers in the field, or by some youth ploughing on the side of the mountain, who hears an unusual humming noise, and sees the swarm dimly whirling by overhead, and, may be, gives chase; or he may simply catch the sound, when he pauses, looks quickly around, but sees nothing. When he comes in at night he tells how he heard or saw a swarm of bees go over; and, perhaps from beneath one of the hives in the garden a black mass of bees has disappeared during the day.

They are not partial as to the kind of tree,—pine, hemlock, elm, birch, maple, hickory,—any tree with a good cavity high up or low down. A swarm of mine ran away from the new patent hive I gave them, and took up their quarters in the hollow trunk of an old apple-tree across an adjoining field. The entrance was a mouse-hole near the ground.

Another swarm in the neighborhood deserted their keeper and went into the cornice of an out-house that stood amid evergreens in the rear of a large mansion. But there is no accounting for the taste of bees, as Samson found when he discovered the swarm in the carcass, or more probably the skeleton, of the lion he had slain.

In any given locality, especially in the more wooded and mountainous districts, the number of swarms that thus assert their independence forms quite a large per cent. In the Northern States these swarms very often perish before spring; but in such a country as Florida they seem to multiply, till bee-trees are very common. In the West, also, wild honey is often gathered in large quantities. I noticed not long since, that some wood-choppers on the west slope of the Coast Range felled a tree that had several pailfuls in it.

One night on the Potomac a party of us unwittingly made our camp near the foot of a bee-tree, which next day the winds of heaven blew down, for our special delectation, at least so we read the sign. Another time while sitting by a waterfall in the leafless April woods I discovered a swarm in the top of a large hickory. I had the season before remarked the tree as a likely place for bees, but the screen of leaves concealed them from me. This time my former presentiment occurred to me, and, looking sharply, sure enough there were the bees, going out and in a large, irregular opening. In June a violent tempest of wind and rain demolished the tree, and the honey was all lost in the creek into which it fell. I happened along that way two or three days after the tornado, when I saw a remnant of the swarm, those, doubtless, that escaped the flood and those that were away when the disaster came, hanging in a small black mass to a branch high up near where their home used to be. They looked forlorn enough. If the queen was saved the remnant probably sought another tree; otherwise the bees have soon died.

I have seen bees desert their hive in the spring when it was infested with worms, or when the honey was exhausted; at such times the swarm seems to wander aimlessly, alighting here and there, and perhaps in the end uniting with some other colony. In case of such union, it would be curious to know if negotiations were first opened between the parties, and if the houseless bees are admitted at once to all the rights and franchises of their benefactors. It would be very like the bees to have some preliminary plan and understanding about the matter on both sides.

Bees will accommodate themselves to almost any quarters, yet no hive seems to please them so well as a section of a hollow tree—"gums" as they are called in the South and West where the sweet gum grows. In some European countries the hive is always made from the trunk of a tree, a suitable cavity being formed by boring. The old-fashioned straw hive is picturesque, and a great favorite with the bees also.

The life of a swarm of bees is like an active and hazardous campaign of an army; the ranks are being continually depleted, and continually recruited. What adventures they have by flood and field, and what hair-breadth escapes! A strong swarm during the honey season loses, on an average, about four or five thousand per month, or one hundred and fifty per day. They are overwhelmed by wind and rain, caught by spiders, benumbed by cold, crushed by cattle, drowned in rivers and ponds, and in many nameless ways cut off or disabled. In the spring the principal mortality is from the cold. As the sun declines they get chilled before they can reach home. Many fall down outside the hive, unable to get in with their burden. One may see them come utterly spent and drop hopelessly into the grass in front of their very doors. Before they can rest the cold has stiffened them. I go out in April and May and pick them up by the handfuls, their baskets loaded with pollen, and warm them in the sun or in the house, or by the simple warmth of my hand, until they can crawl into the hive. Heat is their life, and an apparently lifeless bee may be revived by warming him. I have also picked them up while rowing on the river and seen them safely to shore. It is amusing to see them come hurrying home when there is a thunderstorm approaching. They come piling in till the rain is upon them. Those that are overtaken by the storm doubtless weather it as best they can in the sheltering trees or grass. It is not probable that a bee ever gets lost by wandering into strange and unknown parts. With their myriad eyes they see everything; and then, their sense of locality is very acute, is, indeed, one of their ruling traits. When a bee marks the place of his hive, or of a bit of good pasturage in the fields or swamps, or of the bee-hunter's box of honey on the hills or in the woods, he returns to it as unerringly as fate.

Honey was a much more important article of food with the ancients than it is with us. As they appear to have been unacquainted with sugar, honey, no doubt, stood them instead. It is too rank and pungent for the modern taste; it soon cloys upon the palate. It demands the appetite of youth, and the strong, robust digestion of people who live much in the open air. It is a more wholesome food than sugar, and modern confectionery is poison beside it. Beside grape sugar, honey contains manna, mucilage, pollen, acid, and other vegetable odoriferous substances and juices. It is a sugar with a kind of wild natural bread added. The manna of itself is both food and medicine, and the pungent vegetable extracts have rare virtues. Honey promotes the excretions and dissolves the glutinous and starchy impedimenta of the system.

Hence it is not without reason that with the ancients a land flowing with milk and honey should mean a land abounding in all good things; and the queen in the nursery rhyme, who lingered in the kitchen to eat "bread and honey" while the "king was in the parlor counting out his money," was doing a very sensible thing. Epaminondas is said to have rarely eaten anything but bread and honey. The Emperor Augustus one day inquired of a centenarian how he had kept his vigor of mind and body so long; to which the veteran replied that it was by "oil without and honey within." Cicero, in his "Old Age," classes honey with meat and milk and cheese as among the staple articles with which a well-kept farm-house will be supplied.

Italy and Greece, in fact all the Mediterranean countries, appear to have been famous lands for honey. Mount Hymettus, Mount Hybla, and Mount Ida produced what may be called the classic honey of antiquity, an article doubtless in nowise superior to our best products. Leigh Hunt's "Jar of Honey" is mainly distilled from Sicilian history and literature, Theocritus furnishing the best yield. Sicily has always been rich in bees. Swinburne (the traveler of a hundred years ago) says the woods on this island abounded in wild honey, and that the people also had many hives near their houses. The idyls of Theocritus are native to the island in this respect, and abound in bees—"Flat-nosed bees" as he calls them in the Seventh Idyl—and comparisons in which comb-honey is the standard of the most delectable of this world's goods. His goatherds can think of no greater bliss than that the mouth be filled with honey-combs, or to be inclosed in a chest like Daphnis and fed on the combs of bees; and among the delectables with which Arsinoe cherishes Adonis are "honey-cakes," and other tid-bits made of "sweet honey." In the country of Theocritus this custom is said still to prevail: when a couple are married the attendants place honey in their mouths, by which they would symbolize the hope that their love may be as sweet to their souls as honey to the palate.

It was fabled that Homer was suckled by a priestess whose breasts distilled honey; and that once when Pindar lay asleep the bees dropped honey upon his lips. In the Old Testament the food of the promised Immanuel was to be butter and honey (there is much doubt about the butter in the original), that he might know good from evil; and Jonathan's eyes were enlightened, by partaking of some wood or wild honey: "See, I pray you, how mine eyes have been enlightened, because I tasted a little of this honey." So far as this part of his diet was concerned, therefore, John the Baptist, during his sojourn in the wilderness, his divinity school-days in the mountains and plains of Judea, fared extremely well. About the other part, the locusts, or, not to put too fine a point on it, the grasshoppers, as much cannot be said, though they were among the creeping and leaping things the children of Israel were permitted to eat. They were probably not eaten raw, but roasted in that most primitive of ovens, a hole in the ground made hot by building a fire in it. The locusts and honey may have been served together, as the Bedas of Ceylon are said to season their meat with honey. At any rate, as the locust is often a great plague in Palestine, the prophet in eating them found his account in the general weal, and in the profit of the pastoral bees; the fewer locusts, the more flowers. Owing to its numerous wild-flowers and flowering shrubs, Palestine has always been a famous country for bees. They deposit their honey in hollow trees as our bees do when they escape from the hive, and in holes in the rocks as ours do not. In a tropical or semi-tropical climate bees are quite apt to take refuge in the rocks, but where ice and snow prevail, as with us, they are much safer high up in the trunk of a forest tree.

The best honey is the product of the milder parts of the temperate zone. There are too many rank and poisonous plants in the tropics. Honey from certain districts of Turkey produces headache and vomiting, and that from Brazil is used chiefly as medicine. The honey of Mount Hymettus owes its fine quality to wild thyme. The best honey in Persia and in Florida is collected from the orange blossom. The celebrated honey of Narbonne in the south of France is obtained from a species of rosemary. In Scotland good honey is made from the blossoming heather.

California honey is white and delicate and highly perfumed, and now takes the lead in the market. But honey is honey the world over; and the bee is the bee still. "Men may degenerate," says an old traveler, "may forget the arts by which they acquired renown; manufactories may fail, and commodities be debased, but the sweets of the wild-flowers of the wilderness, the industry and natural mechanics of the bee, will continue without change or derogation."

II. SHARP EYES
AND OTHER PAPERS

SHARP EYES.

Noting how one eye seconds and reinforces the other, I have often amused myself by wondering what the effect would be if one could go on opening eye after eye to the number say of a dozen or more. What would he see? Perhaps not the invisible—not the odors of flowers nor the fever germs in the air—not the infinitely small of the microscope nor the infinitely distant of the telescope. This would require, not more eyes so much as an eye constructed with more and different lenses; but would he not see with augmented power within the natural limits of vision? At any rate some persons seem to have opened more eyes than others, they see with such force and distinctness; their vision penetrates the tangle and obscurity where that of others fails like a spent or impotent bullet. How many eyes did Gilbert White open? how many did Henry Thoreau? how many did Audubon? how many does the hunter, matching his sight against the keen and alert sense of a deer or a moose, or a fox or a wolf? Not outward eyes, but inward. We open another eye whenever we see beyond the first general features or outlines of things—whenever we grasp the special details and characteristic markings that this mask covers. Science confers new powers of vision.

Whenever you have learned to discriminate the birds, or the plants, or the geological features of a country, it is as if new and keener eyes were added.

Of course one must not only see sharply, but read aright what he sees. The facts in the life of Nature that are transpiring about us are like written words that the observer is to arrange into sentences. Or the writing is in cipher and he must furnish the key. A female oriole was one day observed very much preoccupied under a shed where the refuse from the horse stable was thrown. She hopped about among the barn fowls, scolding them sharply when they came too near her. The stable, dark and cavernous, was just beyond. The bird, not finding what she wanted outside, boldly ventured into the stable, and was presently captured by the farmer. What did she want? was the query. What, but a horsehair for her nest which was in an apple-tree near by; and she was so bent on having one that I have no doubt she would have tweaked one out of the horse's tail had he been in the stable. Later in the season I examined her nest and found it sewed through and through with several long horse hairs, so that the bird persisted in her search till the hair was found.

Little dramas and tragedies and comedies, little characteristic scenes, are always being enacted in the lives of the birds, if our eyes are sharp enough to see them. Some clever observer saw this little comedy played among some English sparrows and wrote an account of it in his newspaper; it is too good not to be true: A male bird brought to his box a large, fine goose feather, which is a great find for a sparrow and much coveted. After he had deposited his prize and chattered his gratulations over it he went away in quest of his mate. His next-door neighbor, a female bird, seeing her chance, quickly slipped in and seized the feather,— and here the wit of the bird came out, for instead of carrying it into her own box she flew with it to a near tree and hid it in a fork of the branches, then went home, and when her neighbor returned with his mate was innocently employed about her own affairs. The proud male, finding his feather gone, came out of his box in a high state of excitement, and, with wrath in his manner and accusation on his tongue, rushed into the cot of the female. Not finding his goods and chattels there as he had expected, he stormed around a while, abusing everybody in general and his neighbor in particular, and then went away as if to repair the loss. As soon as he was out of sight, the shrewd thief went and brought the feather home and lined her own domicile with it.

I was much amused one summer day in seeing a bluebird feeding her young one in the shaded street of a large town. She had captured a cicada or harvest-fly, and after bruising it a while on the ground flew with it to a tree and placed it in the beak of the young bird. It was a large morsel, and the mother seemed to have doubts of her chick's ability to dispose of it, for she stood near and watched its efforts with great solicitude. The young bird struggled valiantly with the cicada, but made no head way in swallowing it, when the mother took it from him and flew to the sidewalk, and proceeded to break and bruise it more thoroughly. Then she again placed it in his beak, and seemed to say, "There, try it now," and sympathized so thoroughly with his efforts that she repeated many of his motions and contortions. But the great fly was unyielding, and, indeed, seemed ridiculously disproportioned to the beak that held it. The young bird fluttered and fluttered and screamed, "I'm stuck, I'm stuck," till the anxious parent again seized the morsel and carried it to an iron railing, where she came down upon it for the space of a minute with all the force and momentum her beak could command. Then she offered it to her young a third time, but with the same result as before, except that this time the bird dropped it; but she was at the ground as soon as the cicada was, and taking it in her beak flew some distance to a high board fence where she sat motionless for some moments. While pondering the problem how that fly should be broken, the male bluebird approached her, and said very plainly, and I thought rather curtly, "Give me that bug," but she quickly resented his interference and flew farther away, where she sat apparently quite discouraged when I last saw her.

The bluebird is a home bird, and I am never tired of recurring to him. His coming or reappearance in the spring marks a new chapter in the progress of the season; things are never quite the same after one has heard that note. The past spring the males came about a week in advance of the females. A fine male lingered about my grounds and orchard all the time, apparently waiting the arrival of his mate. He called and warbled every day, as if he felt sure she was within ear-shot, and could be hurried up. Now he warbled half-angrily or upbraidingly, then coaxingly, then cheerily and confidently, the next moment in a plaintive, far-away manner. He would half open his wings, and twinkle them caressingly, as if beckoning his mate to his heart. One morning she had come, but was shy and reserved. The fond male flew to a knot-hole in an old apple-tree, and coaxed her to his side. I heard a fine confidential warble,—the old, old story. But the female flew to a near tree, and uttered her plaintive, homesick note. The male went and got some dry grass or bark in his beak, and flew again to the hole in the old tree, and promised unremitting devotion, but the other said "nay," and flew away in the distance. When he saw her going, or rather heard her distant note, he dropped his stuff, and cried out in a tone that said plainly enough, "Wait a minute. One word, please," and flew swiftly in pursuit. He won her before long, however, and early in April the pair were established in one of the four or five boxes I had put up for them, but not until they had changed their minds several times. As soon as the first brood had flown, and while they were yet under their parents' care, they began another nest in one of the other boxes, the female, as usual, doing all the work, and the male all the complimenting.

A source of occasional great distress to the mother-bird was a white cat that sometimes followed me about. The cat had never been known to catch a bird, but she had a way of watching them that was very embarrassing to the bird. Whenever she appeared, the mother bluebird would set up that pitiful melodious plaint. One morning the cat was standing by me, when the bird came with her beak loaded with building material, and alighted above me to survey the place before going into the box. When she saw the cat, she was greatly disturbed, and in her agitation could not keep her hold upon all her material. Straw after straw came eddying down, till not half her original burden remained. After the cat had gone away, the bird's alarm subsided, till, presently seeing the coast clear, she flew quickly to the box and pitched in her remaining straws with the greatest precipitation, and, without going in to arrange them, as was her wont, flew away in evident relief.

In the cavity of an apple-tree but a few yards off, and much nearer the house than they usually build, a pair of high-holes, or golden-shafted woodpeckers, took up their abode. A knot-hole which led to the decayed interior was enlarged, the live wood being cut away as clean as a squirrel would have done it. The inside preparations I could not witness, but day after day, as I passed near, I heard the bird hammering away, evidently beating down obstructions and shaping and enlarging the cavity. The chips were not brought out, but were used rather to floor the interior. The woodpeckers are not nest-builders, but rather nest-carvers.

The time seemed very short before the voices of the young were heard in the heart of the old tree,—at first feebly, but waxing stronger day by day until they could be heard many rods distant. When I put my hand upon the trunk of the tree, they would set up an eager, expectant chattering; but if I climbed up it toward the opening, they soon detected the unusual sound and would hush quickly, only now and then uttering a warning note. Long before they were fully fledged they clambered up to the orifice to receive their food. As but one could stand in the opening at a time, there was a good deal of elbowing and struggling for this position. It was a very desirable one aside from the advantages it had when food was served; it looked out upon the great shining world, into which the young birds seemed never tired of gazing. The fresh air must have been a consideration also, for the interior of a high-hole's dwelling is not sweet. When the parent birds came with food the young one in the opening did not get it all, but after he had received a portion, either on his own motion or on a hint from the old one, he would give place to the one behind him. Still, one bird evidently outstripped his fellows, and in the race of life, was two or three days in advance of them. His voice was loudest and his head oftenest at the window. But I noticed that when he had kept the position too long, the others evidently made it uncomfortable in his rear, and, after "fidgeting" about a while, he would be compelled to "back down." But retaliation was then easy, and I fear his mates spent few easy moments at that lookout. They would close their eyes and slide back into the cavity as if the world had suddenly lost all its charms for them.

This bird was, of course, the first to leave the nest. For two days before that event he kept his position in the opening most of the time and sent forth his strong voice incessantly. The old ones abstained from feeding him almost entirely, no doubt to encourage his exit. As I stood looking at him one afternoon and noting his progress, he suddenly reached a resolution,—seconded, I have no doubt, from the rear,—and launched forth upon his untried wings. They served him well and carried him about fifty yards up-hill the first heat. The second day after, the next in size and spirit left in the same manner; then another, till only one remained. The parent birds ceased their visits to him, and for one day he called and called till our ears were tired of the sound. His was the faintest heart of all. Then he had none to encourage him from behind. He left the nest and clung to the outer bowl of the tree, and yelped and piped for an hour longer; then he committed himself to his wings and went his way like the rest.

A young farmer in the western part of New York, who has a sharp, discriminating eye, sends me some interesting notes about a tame high-hole he once had.

"Did you ever notice," says he, "that the high-hole never eats anything that he cannot pick up with his tongue? At least this was the case with a young one I took from the nest and tamed. He could thrust out his tongue two or three inches, and it was amusing to see his efforts to eat currants from the hand. He would run out his tongue and try to stick it to the currant; failing in that, he would bend his tongue around it like a hook and try to raise it by a sudden jerk. But he never succeeded, the round fruit would roll and slip away every time. He never seemed to think of taking it in his beak. His tongue was in constant use to find out the nature of everything he saw; a nail-hole in a board or any similar hole was carefully explored. If he was held near the face he would soon be attracted by the eye and thrust his tongue into it. In this way he gained the respect of a number of half-grown cats that were around the house. I wished to make them familiar to each other, so there would be less danger of their killing him. So I would take them both on my knee, when the bird would soon notice the kitten's eyes, and leveling his bill as carefully as a marksman levels his rifle, he would remain so a minute when he would dart his tongue into the cat's eye. This was held by the cats to be very mysterious: being struck in the eye by something invisible to them. They soon acquired such a terror of him that they would avoid him and run away whenever they saw his bill turned in their direction. He never would swallow a grasshopper even when it was placed in his throat; he would shake himself until he had thrown it out of his mouth. His 'best hold' was ants. He never was surprised at anything, and never was afraid of anything. He would drive the turkey gobbler and the rooster. He would advance upon them holding one wing up as high as possible, as if to strike with it, and shuffle along the ground toward them, scolding all the while in a harsh voice. I feared at first that they might kill him, but I soon found that he was able to take care of himself. I would turn over stones and dig into ant-hills for him, and he would lick up the ants so fast that a stream of them seemed going into his mouth unceasingly. I kept him till late in the fall, when he disappeared, probably going south, and I never saw him again."

My correspondent also sends me some interesting observations about the cuckoo. He says a large gooseberry bush standing in the border of an old hedgerow, in the midst of open fields, and not far from his house, was occupied by a pair of cuckoos for two seasons in succession, and, after an interval of a year, for two seasons more. This gave him a good chance to observe them. He says the mother-bird lays a single egg, and sits upon it a number of days before laying the second, so that he has seen one young bird nearly grown, a second just hatched, and a whole egg all in the nest at once. "So far as I have seen, this is the settled practice,—the young leaving the nest one at a time to the number of six or eight. The young have quite the look of the young of the dove in many respects. When nearly grown they are covered with long blue pin-feathers as long as darning-needles, without a bit of plumage on them. They part on the back and hang down on each side by their own weight. With its curious feathers and misshapen body the young bird is anything but handsome. They never open their mouths when approached, as many young birds do, but sit perfectly still, hardly moving when touched." He also notes the unnatural indifference of the mother-bird when her nest and young are approached. She makes no sound, but sits quietly on a near branch in apparent perfect unconcern.

These observations, together with the fact that the egg of the cuckoo is occasionally found in the nests of other birds, raise the inquiry whether our bird is slowly relapsing into the habit of the European species, which always foists its egg upon other birds; or whether, on the other hand, it is not mending its manners in this respect. It has but little to unlearn or to forget in the one case, but great progress to make in the other. How far is its rudimentary nest—a mere platform of coarse twigs and dry stalks of weeds—from the deep, compact, finely woven and finely modeled nest of the goldfinch or king-bird, and what a gulf between its indifference toward its young and their solicitude! Its irregular manner of laying also seems better suited to a parasite like our cow-bird, or the European cuckoo, than to a regular nest-builder.

This observer, like most sharp-eyed persons, sees plenty of interesting things as he goes about his work. He one day saw a white swallow, which is of rare occurrence. He saw a bird, a sparrow he thinks, fly against the side of a horse and fill his beak with hair from the loosened coat of the animal. He saw a shrike pursue a chickadee, when the latter escaped by taking refuge in a small hole in a tree. One day in early spring he saw two hen-hawks that were circling and screaming high in air, approach each other, extend a claw, and, clasping them together, fall toward the earth flapping and struggling as if they were tied together; on nearing the ground they separated and soared aloft again. He supposed that it was not a passage of war but of love, and that the hawks were toying fondly with each other.

He further relates a curious circumstance of finding a humming-bird in the upper part of a barn with its bill stuck fast in a crack of one of the large timbers, dead, of course, with wings extended, and as dry as a chip. The bird seems to have died as it had lived, on the wing, and its last act was indeed a ghastly parody of its living career. Fancy this nimble, flashing sprite, whose life was passed probing the honeyed depths of flowers, at last thrusting its bill into a crack in a dry timber in a hayloft, and, with spread wings, ending its existence.

When the air is damp and heavy, swallows frequently hawk for insects about cattle and moving herds in the field. My farmer describes how they attended him one foggy day, as he was mowing in the meadow with a mowing-machine. It had been foggy for two days, and the swallows were very hungry, and the insects stupid and inert. When the sound of his machine was heard, the swallows appeared and attended him like a brood of hungry chickens. He says there was a continued rush of purple wings over the "cut-bar," and just where it was causing the grass to tremble and fall. Without his assistance the swallows would doubtless have gone hungry yet another day.

Of the hen-hawk, he has observed that both male and female take part in incubation. "I was rather surprised," he says, "on one occasion, to see how quickly they change places on the nest. The nest was in a tall beech, and the leaves were not yet fully out. I could see the head and neck of the hawk over the edge of the nest, when I saw the other hawk coming down through the air at full speed. I expected he would alight near by, but instead of that he struck directly upon the nest, his mate getting out of the way barely in time to avoid being hit; it seemed almost as if he had knocked her off the nest. I hardly see how they can make such a rush on the nest without danger to the eggs."

The king-bird will worry the hawk as a whiffet dog will worry a bear. It is by his persistence and audacity, not by any injury he is capable of dealing his great antagonist. The king-bird seldom more than dogs the hawk, keeping above and between his wings, and making a great ado; but my correspondent says he once "saw a king-bird riding on a hawk's back. The hawk flew as fast as possible, and the king-bird sat upon his shoulders in triumph until they had passed out of sight,"—tweaking his feathers, no doubt, and threatening to scalp him the next moment.

That near relative of the king-bird, the great crested fly-catcher, has one well known peculiarity: he appears never to consider his nest finished until it contains a cast-off snake-skin. My alert correspondent one day saw him eagerly catch up an onion skin and make off with it, either deceived by it or else thinking it a good substitute for the coveted material.

One day in May, walking in the woods, I came upon the nest of a whippoorwill, or rather its eggs, for it builds no nest,—two elliptical whitish spotted eggs lying upon the dry leaves. My foot was within a yard of the mother-bird before she flew. I wondered what a sharp eye would detect curious or characteristic in the ways of the bird, so I came to the place many times and had a look. It was always a task to separate the bird from her surroundings though I stood within a few feet of her, and knew exactly where to look. One had to bear on with his eye, as it were, and refuse to be baffled. The sticks and leaves, and bits of black or dark-brown bark, were all exactly copied in the bird's plumage. And then she did sit so close, and simulate so well a shapeless decaying piece of wood or bark! Twice I brought a companion, and guiding his eye to the spot, noted how difficult it was for him to make out there, in full view upon the dry leaves, any semblance to a bird. When the bird returned after being disturbed, she would alight within a few inches of her eggs, and then, after a moment's pause, hobble awkwardly upon them.

After the young had appeared, all the wit of the bird came into play. I was on hand the next day, I think. The mother-bird sprang up when I was within a pace of her, and in doing so fanned the leaves with her wings till they sprang up too; as the leaves started the young started, and, being of the same color, to tell which was the leaf and which the bird was a trying task to any eye. I came the next day, when the same tactics were repeated. Once a leaf fell upon one of the young birds and nearly hid it. The young are covered with a reddish down like a young partridge, and soon follow their mother about. When disturbed, they gave but one leap, then settled down, perfectly motionless and stupid, with eyes closed. The parent bird, on these occasions made frantic efforts to decoy me away from her young. She would fly a few paces and fall upon her breast, and a spasm, like that of death, would run through her tremulous outstretched wings and prostrate body. She kept a sharp eye out the meanwhile to see if the ruse took, and if it did not, she was quickly cured, and moving about to some other point tried to draw my attention as before. When followed she always alighted upon the ground, dropping down in a sudden peculiar way. The second or third day both old and young had disappeared.

The whippoorwill walks as awkwardly as a swallow, which is as awkward as a man in a bag, and yet she manages to lead her young about the woods. The latter, I think, move by leaps and sudden spurts, their protective coloring shielding them most effectively. Wilson once came upon the mother-bird and her brood in the woods, and, though they were at his very feet, was so baffled by the concealment of the young that he was about to give up the search, much disappointed, when he perceived something "like a slight moldiness among the withered leaves, and, on stooping down, discovered it to be a young whippoorwill seemingly asleep." Wilson's description of the young is very accurate, as its downy covering does look precisely like a "slight moldiness." Returning a few moments afterward to the spot to get a pencil he had forgotten, he could find neither old nor young.

It takes an eye to see a partridge in the woods motionless

upon the leaves; this sense needs to be as sharp as that of smell in hounds and pointers; and yet I know an unkempt youth that seldom fails to see the bird and shoot it before it takes wing. I think he sees it as soon as it sees him and before it suspects itself seen. What a training to the eye is hunting! To pick out the game from its surroundings, the grouse from the leaves, the gray squirrel from the mossy oak limb it hugs so closely, the red fox from the ruddy or brown or gray field, the rabbit from the stubble, or the white hare from the snow requires the best powers of this sense. A woodchuck, motionless in the fields or upon a rock, looks very much like a large stone or bowlder, yet a keen eye knows the difference at a glance, a quarter of a mile away.

A man has a sharper eye than a dog, or a fox, or than any of the wild creatures, but not so sharp an ear or nose. But in the birds he finds his match. How quickly the old turkey discovers the hawk, a mere speck against the sky, and how quickly the hawk discovers you if you happen to be secreted in the bushes or behind the fence near which he alights! One advantage the bird surely has, and that is, owing to the form, structure, and position of the eye, it has a much larger field of vision—indeed, can probably see in nearly every direction at the same instant, behind as well as before. Man's field of vision embraces less than half a circle horizontally, and still less vertically; his brow and brain prevent him from seeing within many degrees of the zenith without a movement of the head; the bird on the other hand, takes in nearly the whole sphere at a glance.

I find I see almost without effort nearly every bird within sight in the field or wood I pass through (a flit of the wing, a flirt of the tail are enough, though the flickering leaves do all conspire to hide them), and that with like ease the birds see me, though, unquestionably, the chances are immensely in their favor. The eye sees what it has the means of seeing, truly. You must have the bird in your heart before you can find it in the bush. The eye must have purpose and aim. No one ever yet found the walking fern who did not have the walking fern in his mind. A person whose eye is full of Indian relics picks them up in every field he walks through.

One season I was interested in the tree-frogs; especially the tiny piper that one hears about the woods and brushy fields— the hyla of the swamps become a denizen of the trees; I had never seen him in this new role. But this season, having hylas in mind, or rather being ripe for them, I several times came across them. One Sunday, walking amid some bushes, I captured two. They leaped before me as doubtless they had done many times before; but though I was not looking for or thinking of them, yet they were quickly recognized, because the eye had been commissioned to find them. On another occasion, not long afterward, I was hurriedly loading my gun in the October woods in hopes of overtaking a gray squirrel that was fast escaping through the tree-tops, when one of these lilliput frogs, the color of the fast-yellowing leaves, leaped near me. I saw him only out of the corner of my eye and yet bagged him, because I had already made him my own.

Nevertheless, the habit of observation is the habit of clear and decisive gazing. Not by a first casual glance, but by a steady deliberate aim of the eye are the rare and characteristic things discovered You must look intently and hold your eye firmly to the spot, to see more than do the rank and file of mankind, The sharp-shooter picks out his man and knows him with fatal certainty from a stump, or a rock, or a cap on a pole. The phrenologists do well to locate, not only form, color, and weight, in the region of the eye, but also a faculty which they call individuality—that which separates, discriminates, and sees in every object its essential character. This is just as necessary to the naturalist as to the artist or the poet. The sharp eye notes specific points and differences,—it seizes upon and preserves the individuality of the thing. Persons frequently describe to me some bird they have seen or heard and ask me to name it, but in most cases the bird might be any one of a dozen, or else it is totally unlike any bird found in this continent. They have either seen falsely or else vaguely. Not so the farm youth who wrote me one winter day that he had seen a single pair of strange birds, which he describes as follows: "They were about the size of the 'chippie,' the tops of their heads were red, and the breast of the male was of the same color, while that of the female was much lighter; their rumps were also faintly tinged with red. If I have described them so that you would know them, please write me their names." There can be little doubt but the young observer had seen a pair of red-polls,—a bird related to the goldfinch, and that occasionally comes down to us in the winter from the far north. Another time, the same youth wrote that he had seen a strange bird, the color of a sparrow, that alighted on fences and buildings as well as upon the ground, and that walked. This last fact shoved the youth's discriminating eye and settled the case. I knew it to be a species of the lark, and from the size, color, season, etc., the tit-lark. But how many persons would have observed that the bird walked instead of hopped?

Some friends of mine who lived in the country tried to describe to me a bird that built a nest in a tree within a few feet of the house. As it was a brown bird, I should have taken it for a wood-thrush, had not the nest been described as so thin and loose that from beneath the eggs could be distinctly seen. The most pronounced feature in the description was the barred appearance of the under side of the bird's tail. I was quite at sea, until one day, when we were driving out, a cuckoo flew across the road in front of us, when my friends exclaimed, "There is our bird!" I had never known a cuckoo to build near a house, and I had never noted the appearance the tail presents when viewed from beneath; but if the bird had been described in its most obvious features, as slender, with a long tail, cinnamon brown above and white beneath, with a curved bill, anyone who knew the bird would have recognized the portrait.

We think we have looked at a thing sharply until we are asked for its specific features. I thought I knew exactly the form of the leaf of the tulip-tree, until one day a lady asked me to draw the outline of one. A good observer is quick to take a hint and to follow it up. Most of the facts of nature, especially in the life of the birds and animals, are well screened. We do not see the play because we do not look intently enough. The other day I was sitting with a friend upon a high rock in the woods, near a small stream, when we saw a water-snake swimming across a pool toward the opposite bank. Any eye would have noted it, perhaps nothing more. A little closer and sharper gaze revealed the fact that the snake bore something in its mouth, which, as we went down to investigate, proved to be a small cat-fish, three or four inches long. The snake had captured it in the pool, and, like any other fisherman, wanted to get its prey to dry land, although itself lived mostly in the water. Here, we said, is being enacted a little tragedy, that would have escaped any but sharp eyes. The snake, which was itself small, had the fish by the throat, the hold of vantage among all creatures, and clung to it with great tenacity. The snake knew that its best tactics was to get upon dry land as soon as possible. It could not swallow its victim alive, and it could not strangle it in the water. For a while it tried to kill its game by holding it up out of the water, but the fish grew heavy, and every few moments its struggles brought down the snake's head. This would not do. Compressing the fish's throat would not shut off its breath under such circumstances, so the wily serpent tried to get ashore with it, and after several attempts succeeded in effecting a landing on a flat rock. But the fish died hard. Cat-fish do not give up the ghost in a hurry. Its throat was becoming congested, but the snake's distended jaws must have ached. It was like a petrified gape. Then the spectators became very curious and close in their scrutiny, and the snake determined to withdraw from the public gaze and finish the business in hand to its own notions. But, when gently but firmly remonstrated with by my friend with his walking-stick, it dropped the fish and retreated in high dudgeon beneath a stone in the bed of the creek. The fish, with a swollen and angry throat, went its way also.

Birds, I say, have wonderfully keen eyes. Throw a fresh bone or a piece of meat upon the snow in winter, and see how soon the crows will discover it and be on hand. If it be near the house or barn, the crow that first discovers it will alight near it, to make sure he is not deceived; then he will go away, and soon return with a companion. The two alight a few yards from the bone, and after some delay, during which the vicinity is sharply scrutinized, one of the crows advances boldly to within a few feet of the coveted prize. Here he pauses, and if no trick is discovered, and the meat be indeed meat, he seizes it and makes off.

One midwinter I cleared away the snow under an apple-tree near the house and scattered some corn there. I had not seen a blue-jay for weeks, yet that very day one found my corn, and after that several came daily and partook of it, holding the kernels under their feet upon the limbs of the trees and pecking them vigorously.

Of course the woodpecker and his kind have sharp eyes; still I was surprised to see how quickly Downy found out some bones that were placed in a convenient place under the shed to be pounded up for the hens. In going out to the barn I often disturbed him making a meal off the bite of meat that still adhered to them.

"Look intently enough at anything," said a poet to me one day, "and you will see something that would otherwise escape you." I thought of the remark as I sat on a stump in an opening of the woods one spring day. I saw a small hawk approaching; he flew to a tall tulip-tree and alighted on a large limb near the top. He eyed me and I eyed him. Then the bird disclosed a trait that was new to me: he hopped along the limb to a small cavity near the trunk, when he thrust in his head and pulled out some small object and fell to eating it. After he had partaken of it for some minutes he put the remainder back in his larder and flew away. I had seen something like feathers eddying slowly down as the hawk ate, and on approaching the spot found the feathers of a sparrow here and there clinging to the bushes beneath the tree. The hawk then—commonly called the chicken hawk—is as provident as a mouse or a squirrel, and lays by a store against a time of need, but I should not have discovered the fact had I not held my eye on him.

An observer of the birds is attracted by any unusual sound or commotion among them. In May or June, when other birds are most vocal, the jay is a silent bird; he goes sneaking about the orchards and the groves as silent as a pickpocket; he is robbing bird's-nests and he is very anxious that nothing should be said about it; but in the fall none so quick and loud to cry "Thief, thief!" as he. One December morning a troop of jays discovered a little screech-owl secreted in the hollow trunk of an old apple-tree near my house. How they found the owl out is a mystery, since it never ventures forth in the light of day; but they did, and proclaimed the fact with great emphasis. I suspect the bluebirds first told them, for these birds are constantly peeping into holes and crannies, both spring and fall. Some unsuspecting bird had probably entered the cavity prospecting for a place for next year's nest, or else looking out a likely place to pass a cold night, and then had rushed out with important news. A boy who should unwittingly venture into a bear's den when Bruin was at home could not be more astonished and alarmed than a bluebird would be on finding itself in the cavity of a decayed tree with an owl. At any rate the bluebirds joined the jays in calling the attention of all whom it might concern to the fact that a culprit of some sort was hiding from the light of day in the old apple-tree. I heard the notes of warning and alarm and approached to within eye-shot. The bluebirds were cautious and hovered about uttering their peculiar twittering calls; but the jays were bolder and took turns looking in at the cavity, and deriding the poor shrinking owl. A jay would alight in the entrance of the hole and flirt and peer and attitudinize, and then flyaway crying "Thief, thief, thief!" at the top of his voice.

I climbed up and peered into the opening, and could just descry the owl clinging to the inside of the tree. I reached in and took him out, giving little heed to the threatening snapping of his beak. He was as red as a fox and as yellow-eyed as a cat. He made no effort to escape, but planted his claws in my forefinger and clung there with a grip that soon grew uncomfortable. I placed him in the loft of an out-house in hopes of getting better acquainted with him. By day he was a very willing prisoner, scarcely moving at all, even when approached and touched with the hand, but looking out upon the world with half-closed, sleepy eyes. But at night what a change; how alert, how wild, how active! He was like another bird; he darted about with wide, fearful eyes, and regarded me like a cornered cat. I opened the window, and swiftly, but as silent as a shadow, he glided out into the congenial darkness, and perhaps, ere this, has revenged himself upon the sleeping jay or bluebird that first betrayed his hiding-place.

THE APPLE.

Lo! sweetened with the summer light,
The full-juiced apple, waxing over-mellow,
Drops in a silent autumn night.—TENNYSON.

Not a little of the sunshine of our northern winters is surely wrapped up in the apple. How could we winter over without it! How is life sweetened by its mild acids! A cellar well filled with apples is more valuable than a chamber filled with flax and wool. So much sound ruddy life to draw upon, to strike one's roots down into, as it were.

Especially to those whose soil of life is inclined to be a little clayey and heavy, is the apple a winter necessity. It is the natural antidote of most of the ills the flesh is heir to. Full of vegetable acids and aromatics, qualities which act as refrigerants and antiseptics, what an enemy it is to jaundice, indigestion, torpidity of liver, etc. It is a gentle spur and tonic to the whole biliary system. Then I have read that it has been found by analysis to contain more phosphorus than any other vegetable. This makes it the proper food of the scholar and the sedentary man; it feeds his brain and it stimulates his liver. Nor is this all. Besides its hygienic properties, the apple is full of sugar and mucilage, which make it highly nutritious. It is said, "The operators of Cornwall, England, consider ripe apples nearly as nourishing as bread, and far more so than potatoes. In the year 1801—which was a year of much scarcity—apples, instead of being converted into cider, were sold to the poor, and the laborers asserted that they could 'stand their work' on baked apples without meat; whereas a potato diet required either meat or some other substantial nutriment. The French and Germans use apples extensively, so do the inhabitants of all European nations. The laborers depend upon them as an article of food, and frequently make a dinner of sliced apples and bread."

Yet the English apple is a tame and insipid affair compared with the intense, sun-colored and sun-steeped fruit our orchards yield. The English have no sweet apple, I am told, the saccharine element apparently being less abundant in vegetable nature in that sour and chilly climate than in our own. It is well known that the European maple yields no sugar, while both our birch and hickory have sweet in their veins. Perhaps this fact accounts for our excessive love of sweets, which may be said to be a national trait.

The Russian apple has a lovely complexion, smooth and transparent, but the Cossack is not yet all eliminated from it. The only one I have seen—the Duchess of Oldenburg—is as beautiful as a Tartar princess, with a distracting odor, but it is the least bit puckery to the taste.

The best thing I know about Chili is not its guano beds, but this fact which I learn from Darwin's "Voyage," namely, that the apple thrives well there. Darwin saw a town there so completely buried in a wood of apple-trees, that its streets were merely paths in an orchard. The tree indeed thrives so well, that large branches cut off in the spring and planted two or three feet deep in the ground send out roots and develop into fine full-bearing trees by the third year. The people know the value of the apple too. They make cider and wine of it and then from the refuse a white and finely flavored spirit; then by another process a sweet treacle is obtained called honey. The children and the pigs eat little or no other food. He does not add that the people are healthy and temperate, but I have no doubt they are. We knew the apple had many virtues, but these Chilians have really opened a deep beneath a deep. We had found out the cider and the spirits, but who guessed the wine and the honey, unless it were the bees? There is a variety in our orchards called the winesap, a doubly liquid name that suggests what might be done with this fruit.

The apple is the commonest and yet the most varied and beautiful of fruits. A dish of them is as becoming to the centre-table in winter as was the vase of flowers in the summer,—a bouquet of spitzenbergs and greenings and northern spies. A rose when it blooms, the apple is a rose when it ripens. It pleases every sense to which it can be addressed, the touch, the smell, the sight, the taste; and when it falls in the still October days it pleases the ear. It is a call to a banquet, it is a signal that the feast is ready. The bough would fain hold it, but it can now assert its independence; it can now live a life of its own.

Daily the stem relaxes its hold, till finally it lets go completely, and down comes the painted sphere with a mellow thump to the earth, towards which it has been nodding so long. It bounds away to seek its bed, to hide under a leaf, or in a tuft of grass. It will now take time to meditate and ripen! What delicious thoughts it has there nestled with its fellows under the fence, turning acid into sugar, and sugar into wine!

How pleasing to the touch! I love to stroke its polished rondure with my hand, to carry it in my pocket on my tramp over the winter hills, or through the early spring woods. You are company, you red-cheeked spitz, or you salmon-fleshed greening! I toy with you; press your face to mine, toss you in the air, roll you on the ground, see you shine out where you lie amid the moss and dry leaves and sticks. You are so alive! You glow like a ruddy flower. You look so animated I almost expect to see you move. I postpone the eating of you, you are so beautiful! How compact; how exquisitely tinted! Stained by the sun and varnished against the rains. An independent vegetable existence, alive and vascular as my own flesh; capable of being wounded, bleeding, wasting away, and almost of repairing damages!

How it resists the cold! holding out almost as long as the red cheeks of the boys do. A frost that destroys the potatoes and other roots only makes the apple more crisp and vigorous; it peeps out from the chance November snows unscathed. When I see the fruit-vender on the street corner stamping his feet and beating his hands to keep them warm, and his naked apples lying exposed to the blasts, I wonder if they do not ache too to clap their hands and enliven their circulation. But they can stand it nearly as long as the vender can.

Noble common fruit, best friend of man and most loved by him, following him like his dog or his cow, wherever he goes. His homestead is not planted till you are planted, your roots intertwine with his; thriving best where he thrives best, loving the limestone and the frost, the plow and the pruning-knife, you are indeed suggestive of hardy, cheerful industry, and a healthy life in the open air. Temperate, chaste fruit! you mean neither luxury nor sloth, neither satiety nor indolence, neither enervating heats nor the Frigid Zones. Uncloying fruit, fruit whose best sauce is the open air, whose finest flavors only he whose taste is sharpened by brisk work or walking knows; winter fruit, when the fire of life burns brightest; fruit always a little hyperborean, leaning towards the cold; bracing, sub-acid, active fruit. I think you must come from the north, you are so frank and honest, so sturdy and appetizing. You are stocky and homely like the northern races. Your quality is Saxon. Surely the fiery and impetuous south is not akin to you. Not spices or olives or the sumptuous liquid fruits, but the grass, the snow, the grains, the coolness is akin to you. I think if I could subsist on you or the like of you, I should never have an intemperate or ignoble thought, never be feverish or despondent. So far as I could absorb or transmute your quality I should be cheerful, continent, equitable, sweet-blooded, long-lived, and should shed warmth and contentment around.

Is there any other fruit that has so much facial expression as the apple? What boy does not more than half believe they can see with that single eye of theirs? Do they not look and nod to him from the bough? The swaar has one look, the rambo another, the spy another. The youth recognizes the seek-no-further buried beneath a dozen other varieties, the moment he catches a glance of its eye, or the bonny-cheeked Newtown pippin, or the gentle but sharp-nosed gilliflower. He goes to the great bin in the cellar and sinks his shafts here and there in the garnered wealth of the orchards, mining for his favorites, sometimes coming plump upon them, sometimes catching a glimpse of them to the right or left, or uncovering them as keystones in an arch made up of many varieties. In the dark he can usually tell them by the sense of touch. There is not only the size and shape, but there is the texture and polish. Some apples are coarse grained and some are fine; some are thin-skinned and some are thick. One variety is quick and vigorous beneath the touch; another gentle and yielding. The pinnock has a thick skin with a spongy lining, a bruise in it becomes like a piece of cork. The tallow apple has an unctuous feel, as its name suggests. It sheds water like a duck. What apple is that with a fat curved stem that blends so prettily with its own flesh,—the wine-apple? Some varieties impress me as masculine,—weather-stained, freckled, lasting and rugged; others are indeed lady apples, fair, delicate, shining, mild-flavored, white-meated, like the egg-drop and the lady-finger. The practiced hand knows each kind by the touch. Do you remember the apple hole in the garden or back of the house, Ben Bolt? In the fall after the bins in the cellar had been well stocked, we excavated a circular pit in the warm, mellow earth, and covering the bottom with clean rye straw, emptied in basketful after basketful of hardy choice varieties, till there was a tent-shaped mound several feet high of shining variegated fruit. Then wrapping it about with a thick layer of long rye straw, and tucking it up snug and warm, the mound was covered, with a thin coating of earth, a flat stone on the top holding down the straw. As winter set in, another coating of earth was put upon it, with perhaps an overcoat of coarse dry stable manure, and the precious pile was left in silence and darkness till spring. No marmot hibernating under-ground in his nest of leaves and dry grass, more cosy and warm. No frost, no wet, but fragrant privacy and quiet. Then how the earth tempers and flavors the apples! It draws out all the acrid unripe qualities, and infuses into them a subtle refreshing taste of the soil. Some varieties perish; but the ranker, hardier kinds, like the northern spy, the greening, or the black apple, or the russet, or the pinnock, how they ripen and grow in grace, how the green becomes gold, and the bitter becomes sweet!

As the supply in the bins and barrels gets low and spring approaches, the buried treasures in the garden are remembered. With spade and axe we go out and penetrate through the snow and frozen earth till the inner dressing of straw is laid bare. It is not quite as clear and bright as when we placed it there last fall, but the fruit beneath, which the hand soon exposes, is just as bright and far more luscious. Then, as day after day you resort to the hole, and, removing the straw and earth from the opening, thrust your arm into the fragrant pit, you have a better chance than ever before to become acquainted with your favorites by the sense of touch. How you feel for them, reaching to the right and left! Now you have got a Tolman sweet; you imagine you can feel that single meridian line that divides it into two hemispheres. Now a greening fills your hand, you feel its fine quality beneath its rough coat. Now you have hooked a swaar, you recognize its full face; now a Vandevere or a King rolls down from the apex above, and you bag it at once. When you were a school-boy you stowed these away in your pockets and ate them along the road and at recess, and again at noon time; and they, in a measure, corrected the effects of the cake and pie with which your indulgent mother filled your lunch-basket.

The boy is indeed the true apple-eater, and is not to be questioned how he came by the fruit with which his pockets are filled. It belongs to him...His own juicy flesh craves the juicy flesh of the apple. Sap draws sap. His fruit-eating has little reference to the state of his appetite. Whether he be full of meat or empty of meat he wants the apple just the same. Before meal or after meal it never comes amiss. The farm-boy munches apples all day long. He has nests of them in the hay-mow, mellowing, to which he makes frequent visits. Sometimes old Brindle, having access through the open door, smells them out and makes short work of them.

In some countries the custom remains of placing a rosy apple in the hand of the dead that they may find it when they enter paradise. In northern mythology the giants eat apples to keep off old age.

The apple is indeed the fruit of youth. As we grow old we crave apples less. It is an ominous sign. When you are ashamed to be seen eating them on the street; when you can carry them in your pocket and your hand not constantly find its way to them; when your neighbor has apples and you have none, and you make no nocturnal visits to his orchard; when your lunch-basket is without them, and you can pass a winter's night by the fireside with no thought of the fruit at your elbow, then be assured you are no longer a boy, either in heart or years.

The genuine apple-eater comforts himself with an apple in their season as others with a pipe or cigar. When he has nothing else to do, or is bored, he eats an apple. While he is waiting for the train he eats an apple, sometimes several of them. When he takes a walk, he arms himself with apples. His traveling bag is full of apples. He offers an apple to his companion, and takes one himself. They are his chief solace when on the road. He sows their seed all along the route. He tosses the core from the car-window and from the top of the stage-coach. He would, in time, make the land one vast orchard. He dispenses with a knife. He prefers that his teeth shall have the first taste. Then he knows the best flavor is immediately beneath the skin, and that in a pared apple this is lost. If you will stew the apple, he says, instead of baking it, by all means leave the skin on. It improves the color and vastly heightens the flavor of the dish.

The apple is a masculine fruit; hence women are poor apple-eaters. It belongs to the open air, and requires an open-air taste and relish.

I instantly sympathized with that clergyman I read of, who on pulling out his pocket-handkerchief in the midst of his discourse, pulled out two bouncing apples with it that went rolling across the pulpit floor and down the pulpit stairs. These apples were, no doubt, to be eaten after the sermon on his way home, or to his next appointment. They would take the taste of it out of his mouth. Then, would a minister be apt to grow tiresome with two big apples in his coat-tail pockets? Would he not naturally hasten along to "lastly," and the big apples? If they were the dominie apples, and it was April or May, he certainly....

How the early settlers prized the apple! When their trees broke down or were split asunder by the storms, the neighbors turned out, the divided tree was put together again and fastened with iron bolts. In some of the oldest orchards one may still occasionally see a large dilapidated tree with the rusty iron bolt yet visible. Poor, sour fruit, too, but sweet in those early pioneer days. My grandfather, who was one of these heroes of the stump, used every fall to make a journey of forty miles for a few apples, which he brought home in a bag on horseback. He frequently started from home by two or three o'clock in the morning, and at one time both he and his horse were much frightened by the screaming of panthers in a narrow pass in the mountains through which the road led.

Emerson, I believe, has spoken of the apple as the social fruit of New England. Indeed, what a promoter or abettor of social intercourse among our rural population the apple has been, the company growing more merry and unrestrained as soon as the basket of apples was passed round! When the cider followed, the introduction and good understanding were complete. Then those rural gatherings that enlivened the autumn in the country, known as "apple cuts," now, alas! nearly obsolete, where so many things were cut and dried besides apples! The larger and more loaded the orchard, the more frequently the invitations went round and the higher the social and convivial spirit ran. Ours is eminently a country of the orchard. Horace Greeley said he had seen no land in which the orchard formed such a prominent feature in the rural and agricultural districts. Nearly every farmhouse in the Eastern and Northern States has its setting or its background of apple-trees, which generally date back to the first settlement of the farm. Indeed, the orchard, more than almost any other thing, tends to soften and humanize the country, and to give the place of which it is an adjunct, a settled, domestic look. The apple-tree takes the rawness and wildness off any scene. On the top of a mountain, or in remote pastures, it sheds the sentiment of home. It never loses its domestic air, or lapses into a wild state. And in planting a homestead, or in choosing a building site for the new house, what a help it is to have a few old, maternal apple-trees near by; regular old grandmothers, who have seen trouble, who have been sad and glad through so many winters and summers, who have blossomed till the air about them is sweeter than elsewhere, and borne fruit till the grass beneath them has become thick and soft from human contact, and who have nourished robins and finches in their branches till they have a tender, brooding look. The ground, the turf, the atmosphere of an old orchard, seem several stages nearer to man than that of the adjoining field, as if the trees had given back to the soil more than they had taken from it; as if they had tempered the elements and attracted all the genial and beneficent influences in the landscape around.

An apple orchard is sure to bear you several crops beside the apple. There is the crop of sweet and tender reminiscences dating from childhood and spanning the seasons from May to October, and making the orchard a sort of outlying part of the household. You have played there as a child, mused there as a youth or lover, strolled there as a thoughtful, sad-eyed man. Your father, perhaps, planted the trees, or reared them from the seed, and you yourself have pruned and grafted them, and worked among them, till every separate tree has a peculiar history and meaning in your mind. Then there is the never-failing crop of birds— robins, goldfinches, king-birds, cedar-birds, hair-birds, orioles, starlings—all nesting and breeding in its branches, and fitly described by Wilson Flagg as "Birds of the Garden and Orchard." Whether the pippin and sweetbough bear or not, the "punctual birds" can always be depended on. Indeed, there are few better places to study ornithology than in the orchard. Besides its regular occupants, many of the birds of the deeper forest find occasion to visit it during the season. The cuckoo comes for the tent-caterpillar, the jay for frozen apples, the ruffed grouse for buds, the crow foraging for birds' eggs, the woodpecker and chickadees for their food, and the high-hole for ants. The redbird comes too, if only to see what a friendly covert its branches form; and the wood-thrush now and then comes out of the grove near by, and nests alongside of its cousin, the robin. The smaller hawks know that this is a most likely spot for their prey; and in spring the shy northern warblers may be studied as they pause to feed on the fine insects amid its branches. The mice love to dwell here also, and hither comes from the near woods the squirrel and the rabbit. The latter will put his head through the boy's slipper-noose any time for taste of the sweet apple, and the red squirrel and chipmunk esteem its seeds a great rarity.

All the domestic animals love the apple, but none so much so as the cow. The taste of it wakes her up as few other things do, and bars and fences must be well looked after. No need to assort them or pick out the ripe ones for her. An apple is an apple, and there is no best about it. I heard of a quick-witted old cow that learned to shake them down from the tree. While rubbing herself she had observed that an apple sometimes fell. This stimulated her to rub a little harder, when more apples fell. She then took the hint and rubbed her shoulder with such vigor that the farmer had to check her and keep an eye on her to save his fruit.

But the cow is the friend of the apple. How many trees she has planted about the farm, in the edge of the woods, and in remote fields and pastures. The wild apples, celebrated by Thoreau, are mostly of her planting. She browses them down to be sure, but they are hers, and why should she not?

What an individuality the apple-tree has, each variety being nearly as marked by its form as by its fruit. What a vigorous grower, for instance, is the Ribston pippin, an English apple. Wide branching like the oak, and its large ridgy fruit, in late fall or early winter, is one of my favorites. Or the thick and more pendent top of the belleflower, with its equally rich, sprightly uncloying fruit.

Sweet apples are perhaps the most nutritious, and when baked are a feast in themselves. With a tree of the Jersey sweet or of Tolman's sweeting in bearing, no man's table need be devoid of luxuries and one of the most wholesome of all deserts. Or the red astrachan, an August apple, what a gap may be filled in the culinary department of a household at this season, by a single tree of this fruit! And what a feast is its shining crimson coat to the eye before its snow-white flesh has reached the tongue. But the apple of apples for the household is the spitzenberg. In this casket Pomona has put her highest flavors. It can stand the ordeal of cooking and still remain a spitz. I recently saw a barrel of these apples from the orchard of a fruit-grower in the northern part of New York, who has devoted special attention to this variety. They were perfect gems. Not large, that had not been the aim, but small, fair, uniform, and red to the core. How intense, how spicy and aromatic!

But all the excellences of the apple are not confined to the cultivated fruit. Occasionally a seedling springs up about the farm that produces fruit of rare beauty and worth. In sections peculiarly adapted to the apple, like a certain belt along the Hudson River, I have noticed that most of the wild unbidden trees bear good, edible fruit. In cold and ungenial districts, the seedlings are mostly sour and crabbed, but in more favorable soils they are oftener mild and sweet. I know wild apples that ripen in August, and that do not need, if it could be had, Thoreau's sauce of sharp November air to be eaten with. At the foot of a hill near me and striking its roots deep in the shale, is a giant specimen of native tree that bears an apple that has about the clearest, waxiest, most transparent complexion I ever saw. It is good size, and the color of a tea-rose. Its quality is best appreciated in the kitchen. I know another seedling of excellent quality and so remarkable for its firmness and density, that it is known on the farm where it grows as the "heavy apple."

I have alluded to Thoreau, to whom all lovers of the apple and its tree are under obligation. His chapter on Wild Apples is a most delicious piece of writing. It has a "tang and smack" like the fruit it celebrates, and is dashed and streaked with color in the same manner. It has the hue and perfume of the crab, and the richness and raciness of the pippin. But Thoreau loved other apples than the wild sorts and was obliged to confess that his favorites could not be eaten in-doors. Late in November he found a blue-pearmain tree growing within the edge of a swamp, almost as good as wild. "You would not suppose," he says, "that there was any fruit left there on the first survey, but you must look according to system. Those which lie exposed are quite brown and rotten now, or perchance a few still show one blooming cheek here and there amid the wet leaves. Nevertheless, with experienced eyes I explore amid the bare alders, and the huckleberry bushes, and the withered sedge, and in the crevices of the rocks, which are full of leaves, and pry under the fallen and decayed ferns which, with apple and alder leaves, thickly strew the ground. For I know that they lie concealed, fallen into hollows long since, and covered up by the leaves of the tree itself—a proper kind of packing. From these lurking places, everywhere within the circumference of the tree, I draw forth the fruit all wet and glossy, maybe nibbled by rabbits and hollowed out by crickets, and perhaps a leaf or two cemented to it (as Curzon an old manuscript from a monastery's mouldy cellar), but still with a rich bloom on it, and at least as ripe and well kept, if no better than those in barrels, more crisp and lively than they. If these resources fail to yield anything, I have learned to look between the leaves of the suckers which spring thickly from some horizontal limb, for now and then one lodges there, or in the very midst of an alder-clump, where they are covered by leaves, safe from cows which may have smelled them out. If I am sharp-set, for I do not refuse the blue-pearmain, I fill my pockets on each side; and as I retrace my steps, in the frosty eve being perhaps four or five miles from home, I eat one first from this side, and then from that, to keep my balance."

A TASTE OF MAINE BIRCH.

The traveler and camper-out in Maine, unless he penetrates its more northern portions, has less reason to remember it as a pine-tree State than a birch-tree State. The white-pine forests have melted away like snow in the spring and gone down stream, leaving only patches here and there in the more remote and inaccessible parts. The portion of the State I saw—the valley of the Kennebec and the woods about Moxie Lake—had been shorn of its pine timber more than forty years before, and is now covered with a thick growth of spruce and cedar and various deciduous trees. But the birch abounds. Indeed, when the pine goes out the birch comes in; the race of men succeeds the race of giants. This tree has great stay-at-home virtues. Let the sombre, aspiring, mysterious pine go; the birch has humble every-day uses. In Maine, the paper or canoe birch is turned to more account than any other tree. I read in Gibbon that the natives of ancient Assyria used to celebrate in verse or prose the three hundred and sixty uses to which the various parts and products of the palm-tree were applied. The Maine birch is turned to so many accounts that it may well be called the palm of this region. Uncle Nathan, our guide, said it was made especially for the camper-out; yes, and for the wood-man and frontiersman generally. It is a magazine, a furnishing store set up in the wilderness, whose goods are free to every comer. The whole equipment of the camp lies folded in it, and comes forth at the beck of the woodman's axe; tent, waterproof roof, boat, camp utensils, buckets, cups, plates, spoons, napkins, table cloths, paper for letters or your journal, torches, candles, kindling-wood, and fuel. The canoe-birch yields you its vestments with the utmost liberality. Ask for its coat, and it gives you its waistcoat also. Its bark seems wrapped about it layer upon layer, and comes off with great ease. We saw

117

many rude structures and cabins shingled and sided with it, and haystacks capped with it. Near a maple-sugar camp there was a large pile of birch-bark sap-buckets,—each bucket made of a piece of bark about a yard square, folded up as the tinman folds up a sheet of tin to make a square vessel, the corners bent around against the sides and held by a wooden pin. When, one day, we were overtaken by a shower in traveling through the woods, our guide quickly stripped large sheets of the bark from a near tree, and we had each a perfect umbrella as by magic. When the rain was over, and we moved on, I wrapped mine about me like a large leather apron, and it shielded my clothes from the wet bushes. When we came to a spring, Uncle Nathan would have a birch-bark cup ready before any of us could get a tin one out of his knapsack, and I think water never tasted so sweet as from one of these bark cups. It is exactly the thing. It just fits the mouth and it seems to give new virtues to the water. It makes me thirsty now when I think of it. In our camp at Moxie we made a large birch-bark box to keep the butter in; and the butter in this box, covered with some leafy boughs, I think improved in flavor day by day. Maine butter needs something to mollify and sweeten it a little, and I think birch bark will do it. In camp Uncle Nathan often drank his tea and coffee from a bark cup; the china closet in the birch-tree was always handy, and our vulgar tin ware was generally a good deal mixed, and the kitchen-maid not at all particular about dish-washing. We all tried the oatmeal with the maple syrup in one of these dishes, and the stewed mountain cranberries, using a birch-bark spoon, and never found service better. Uncle Nathan declared he could boil potatoes in a bark kettle, and I did not doubt him. Instead of sending our soiled napkins and table-spreads to the wash, we rolled them up into candles and torches, and drew daily upon our stores in the forest for new ones.

But the great triumph of the birch is of course the bark canoe. When Uncle Nathan took us out under his little wood-shed, and showed us, or rather modestly permitted us to see, his nearly finished canoe, it was like a first glimpse of some new and unknown genius of the woods or streams. It sat there on the chips and shavings and fragments of bark like some shy delicate creature just emerged from its hiding-place, or like some wild flower just opened. It was the first boat of the kind I had ever seen, and it filled my eye completely. What woodcraft it indicated, and what a wild free life, sylvan life, it promised! It had such a fresh, aboriginal look as I had never before seen in any kind of handiwork. Its clear yellow-red color would have become the cheek of an Indian maiden. Then its supple curves and swells, its sinewy stays and thwarts, its bow-like contour, its tomahawk stem and stern rising quickly and sharply from its frame, were all vividly suggestive of the race from which it came. An old Indian had taught Uncle Nathan the art, and the soul of the ideal red man looked out of the boat before us. Uncle Nathan had spent two days ranging the mountains looking for a suitable tree, and had worked nearly a week on the craft. It was twelve feet long, and would seat and carry five men nicely. Three trees contribute to the making of a canoe besides the birch, namely, the white cedar for ribs and lining, the spruce for roots and fibres to sew its joints and bind its frame, and the pine for pitch or rosin to stop its seams and cracks. It is hand-made and home-made, or rather wood-made, in a sense that no other craft is, except a dug-out, and it suggests a taste and a refinement that few products of civilization realize. The design of a savage, it yet looks like the thought of a poet, and its grace and fitness haunt the imagination. I suppose its production was the inevitable result of the Indian's wants and surroundings, but that does not detract from its beauty. It is, indeed, one of the fairest flowers the thorny plant of necessity ever bore. Our canoe, as I have intimated, was not yet finished when we first saw it, nor yet when we took it up, with its architect, upon our metaphorical backs and bore it to the woods. It lacked part of its cedar lining and the rosin upon its joints, and these were added after we reached our destination.

Though we were not indebted to the birch-tree for our

guide, Uncle Nathan, as he was known in all the country, yet he matched well these woodsy products and conveniences. The birch-tree had given him a large part of his tuition, and kneeling in his canoe and making it shoot noiselessly over the water with that subtle yet indescribably expressive and athletic play of the muscles of the back and shoulders, the boat and the man seemed born of the same spirit. He had been a hunter and trapper for over forty years; he had grown gray in the woods, had ripened and matured there, and everything about him was as if the spirit of the woods had had the ordering of it; his whole make-up was in a minor and subdued key, like the moss and the lichens, or like the protective coloring of the game,—everything but his quick sense and penetrative glance. He was as gentle and modest as a girl; his sensibilities were like plants that grow in the shade. The woods and the solitudes had touched him with their own softening and refining influence; had indeed shed upon his soil of life a rich deep leaf mould that was delightful, and that nursed, half concealed, the tenderest and wildest growths. There was grit enough back of and beneath it all, but he presented none of the rough and repelling traits of character of the conventional backwoods-man. In the spring he was a driver of logs on the Kennebec, usually having charge of a large gang of men; in the winter he was a solitary trapper and hunter in the forests.

Our first glimpse of Maine waters was Pleasant Pond, which we found by following a white, rapid, musical stream from the Kennebec three miles back into the mountains. Maine waters are for the most part dark-complexioned, Indian-colored streams, but Pleasant Pond is a pale-face among them both in name and nature. It is the only strictly silver lake I ever saw. Its waters seem almost artificially white and brilliant, though of remarkable transparency. I think I detected minute shining motes held in suspension in it. As for the trout they are veritable bars of silver until you have cut their flesh, when they are the reddest of gold. They have no crimson or other spots, and the straight lateral line is but a faint pencil mark. They appeared to be a species of lake trout peculiar to these waters, uniformly from ten to twelve inches in length. And these beautiful fish, at the time of our visit (last of August) at least, were to be taken only in deep water upon a hook baited with salt pork. And then you needed a letter of introduction to them. They were not to be tempted or cajoled by strangers. We did not succeed in raising a fish, although instructed how it was to be done, until one of the natives, a young and obliging farmer living hard by, came and lent his countenance to the enterprise. I sat in one end of the boat and he in the other; my pork was the same as his, and I maneuvered it as directed, and yet those fish knew his hook from mine in sixty feet of water, and preferred it four times in five. Evidently they did not bite because they were hungry, but solely for old acquaintance' sake.

Pleasant Pond is an irregular sheet of water, two miles or more in its greatest diameter, with high, rugged mountains rising up from its western shore, and low rolling hills sweeping back from its eastern and northern, covered by a few sterile farms. I was never tired, when the wind was still, of floating along its margin and gazing down into its marvelously translucent depths. The boulders and fragments of rocks were seen, at a depth of twenty-five or thirty feet, strewing its floor, and apparently as free from any covering of sediment as when they were dropped there by the old glaciers aeons ago. Our camp was amid a dense grove of second growth of white pine on the eastern shore, where, for one, I found a most admirable cradle in a little depression, outside of the tent, carpeted with pine needles, in which to pass the night. The camper-out is always in luck if he can find, sheltered by the trees, a soft hole in the ground, even if he has a stone for a pillow. The earth must open its arms a little for us even in life, if we are to sleep well upon its bosom. I have often heard my grand-father, who was a soldier of the Revolution, tell with great gusto how he once bivouacked in a little hollow made by the overturning of a tree, and slept so soundly that he did not wake up till his cradle was half full of water from a passing shower.

What bird or other creature might represent the divinity of Pleasant Pond I do not know, but its demon, as of most northern inland waters, is the loon, and a very good demon he is too, suggesting something not so much malevolent, as arch, sardonic, ubiquitous, circumventing, with just a tinge of something inhuman and uncanny. His fiery red eyes gleaming forth from that jet-black head are full of meaning. Then his strange horse laughter by day and his weird, doleful cry at night, like that of a lost and wandering spirit, recall no other bird or beast. He suggests something almost supernatural in his alertness and amazing quickness, cheating the shot and the bullet of the sportsman out of their aim. I know of but one other bird so quick, and that is the humming-bird, which I have never been able to kill with a gun. The loon laughs the shot-gun to scorn, and the obliging young farmer above referred to told me he had shot at them hundreds of times with his rifle, without effect,— they always dodged his bullet. We had in our party a breach-loading rifle, which weapon is perhaps an appreciable moment of time quicker than the ordinary muzzleloader, and this the poor loon could not or did not dodge. He had not timed himself to that species of fire-arm, and when, with his fellow, he swam about within rifle range of our camp, letting off volleys of his wild ironical ha-ha, he little suspected the dangerous gun that was matched against him. As the rifle cracked both loons made the gesture of diving, but only one of them disappeared beneath the water; and when he came to the surface in a few moments, a hundred or more yards away, and saw his companion did not follow, but was floating on the water where he had last seen him, he took the alarm and sped away in the distance. The bird I had killed was a magnificent specimen, and I looked him over with great interest. His glossy checkered coat, his banded neck, his snow-white breast, his powerful lance-shaped beak, his red eyes, his black, thin, slender, marvelously delicate feet and legs, issuing from his muscular thighs, and looking as if they had never touched the ground, his strong wings well forward while his legs were quite at the apex, and the neat, elegant model of the entire bird, speed and quickness and strength stamped upon every feature,—all delighted and lingered in the eye. The loon appears like anything but a silly bird, unless you see him in some collection, or in the shop of the taxidermist, where he usually looks very tame and goose-like. Nature never meant the loon to stand up, or to use his feet and legs for other purposes than swimming. Indeed, he cannot stand except upon his tail in a perpendicular attitude, but in the collections he is poised upon his feet like a barn-yard fowl, all the wildness and grace and alertness goes out of him. My specimen sits upon a table as upon the surface of the water, his feet trailing behind him, his body low and trim, his head elevated and slightly turned as if in the act of bringing that fiery eye to bear upon you, and vigilance and power stamped upon every lineament.

The loon is to the fishes what the hawk is to the birds; he swoops down to unknown depths upon them, and not even the wary trout can elude him. Uncle Nathan said he had seen the loon disappear and in a moment come up with a large trout, which he would cut in two with his strong beak, and swallow piecemeal. Neither the loon nor the otter can bolt a fish under the water; he must come to the surface to dispose of it. (I once saw a man eat a cake under water in London.) Our guide told me he had seen the parent loon swimming with a single young one upon its back. When closely pressed it dove, or "div" as he would have it, and left the young bird sitting upon the water. Then it too disappeared, and when the old one returned and called, it came out from the shore. On the wing overhead, the loon looks not unlike a very large duck, but when it alights it ploughs into the water like a bombshell. It probably cannot take flight from the land, as the one Gilbert White saw and describes in his letters was picked up in a field, unable to launch itself into the air.

From Pleasant Pond we went seven miles through the woods to Moxie Lake, following an overgrown lumberman's "tote" road, our canoe and supplies, etc., hauled on a sled by the young farmer with his three-year-old steers. I doubt if birch-bark ever made rougher voyage than that. As I watched it above the bushes, the sled and the luggage being hidden, it appeared as if tossed in the wildest and most tempestuous sea. When the bushes closed above it I felt as if it had gone down, or been broken into a hundred pieces. Billows of rocks and logs, and chasms of creeks and spring runs, kept it rearing and pitching in the most frightful manner. The steers went at a spanking pace; indeed, it was a regular bovine gale; but their driver clung to their side amid the brush and boulders with desperate tenacity, and seemed to manage them by signs and nudges, for he hardly uttered his orders aloud. But we got through without any serious mishap, passing Mosquito Creek and Mosquito Pond, and flanking Mosquito Mountain, but seeing no mosquitoes, and brought up at dusk at a lumberman's old hay-barn, standing in the midst of a lonely clearing on the shores of Moxie Lake.

Here we passed the night, and were lucky in having a good roof over our heads, for it rained heavily. After we were rolled in our blankets and variously disposed upon the haymow, Uncle Nathan lulled us to sleep by a long and characteristic yarn.

I had asked him, half jocosely, if he believed in "spooks"; but he took my question seriously, and without answering it directly, proceeded to tell us what he himself had known and witnessed. It was, by the way, extremely difficult either to surprise or to steal upon any of Uncle Nathan's private opinions and beliefs about matters and things. He was as shy of all debatable subjects as a fox is of a trap. He usually talked in a circle, just as he hunted moose and caribou, so as not to approach his point too rudely and suddenly. He would keep on the lee side of his interlocutor in spite of all one could do. He was thoroughly good and reliable, but the wild creatures of the woods, in pursuit of which he had spent so much of his life, had taught him a curious gentleness and indirection, and to keep himself in the back-ground; he was careful that you should not scent his opinions upon any subject at all polemic, but he would tell you what he had seen and known.

What he had seen and known about spooks was briefly this:—In company with a neighbor he was passing the night with an old recluse who lived somewhere in these woods. Their host was an Englishman, who had the reputation of having murdered his wife some years before in another part of the country, and, deserted by his grown-up children, was eking out his days in poverty amid these solitudes. The three men were sleeping upon the floor, with Uncle Nathan next to a rude partition that divided the cabin into two rooms. At his head there was a door that opened into this other apartment. Late at night, Uncle Nathan said, he awoke and turned over, and his mind was occupied with various things, when he heard somebody behind the partition. He reached over and felt that both of his companions were in their places beside him, and he was somewhat surprised. The person, or whatever it was, in the other room moved about heavily, and pulled the table from its place beside the wall to the middle of the floor. "I was not dreaming," said Uncle Nathan; "I felt of my eyes twice to make sure, and they were wide open." Presently the door opened; he was sensible of the draught upon his head, and a woman's form stepped heavily past him; he felt the "swirl" of her skirts as she went by. Then there was a loud noise in the room as if some one had fallen their whole length upon the floor. "It jarred the house," said he, "and woke everybody up. I asked old Mr. ——— if he heard that noise. 'Yes,' said he, 'it was thunder.' But it was not thunder, I know that;" and then added, "I was no more afraid than I am this minute. I never was the least mite afraid in my life. And my eyes were wide open," he repeated; "I felt of them twice; but whether that was the speret of that man's murdered wife or not I cannot tell. They said she was an uncommon heavy woman." Uncle Nathan was a man of unusually quick and acute senses, and he did not doubt their evidence on this occasion any more than he did when they prompted him to level his rifle at a bear or a moose.

Moxie Lake lies much lower than Pleasant Pond, and its waters compared with those of the latter are as copper compared with silver. It is very irregular in shape; now narrowing to the dimensions of a slow moving grassy creek, then expanding into a broad deep basin with rocky shores, and commanding the noblest mountain scenery. It is rarely that the pond-lily and the speckled trout are found together,—the fish the soul of the purest spring water, the flower the transfigured spirit of the dark mud and slime of sluggish summer streams and ponds; yet in Moxie they were both found in perfection. Our camp was amid the birches, poplars, and white cedars near the head of the lake, where the best fishing at this season was to be had. Moxie has a small oval head, rather shallow, but bumpy with rocks; a long, deep neck, full of springs, where the trout lie; and a very broad chest, with two islands tufted with pine-trees for breasts. We swam in the head, we fished in the neck, or in a small section of it, a space about the size of the Adam's apple, and we paddled across and around the broad expanse below. Our birch bark was not finished and christened till we reached Moxie. The cedar lining was completed at Pleasant Pond, where we had the use of a bateau, but the rosin was not applied to the seams till we reached this lake. When I knelt down in it for the first time and put its slender maple paddle into the water, it sprang away with such quickness and speed that it disturbed me in my seat. I had spurred a more restive and spirited steed than I was used to. In fact, I had never been in a craft that sustained so close a relation to my will, and was so responsive to my slightest wish. When I caught my first large trout from it, it sympathized a little too closely, and my enthusiasm started a leak, which, however, with a live coal and a piece of rosin, was quickly ended. You cannot perform much of a war-dance in a birch-bark canoe: better wait till you get on dry land. Yet as a boat it is not so shy and "ticklish" as I had imagined. One needs to be on the alert, as becomes a sportsman and an angler, and in his dealings with it must charge himself with three things,—precision, moderation, and circumspection.

Trout weighing four and five pounds have been taken at

Moxie, but none of that size came to our hand. I realized the fondest hopes I had dared to indulge in when I hooked the first two-pounder of my life, and my extreme solicitude lest he get away I trust was pardonable. My friend, in relating the episode in camp, said I implored him to row me down in the middle of the lake that I might have room to manoeuver my fish. But the slander has barely a grain of truth in it. The water near us showed several old stakes broken off just below the surface, and my fish was determined to wrap my leader about one of these stakes; it was only for the clear space a few yards farther out that I prayed. It was not long after that my friend found himself in an anxious frame of mind. He hooked a large trout, which came home on him so suddenly that he had not time to reel up his line, and in his extremity he stretched his tall form into the air and lifted up his pole to an incredible height. He checked the trout before it got under the boat, but dared not come down an inch, and then began his amusing further elongation in reaching for his reel with one hand while he carried it ten feet into the air with the other. A step-ladder would perhaps have been more welcome to him just then than at any other moment during his life. But the trout was saved, though my friend's buttons and suspenders suffered.

We learned a new trick in fly-fishing here, worth disclosing. It was not one day in four that the trout would take the fly on the surface. When the south wind was blowing and the clouds threatened rain, they would at times, notably about three o'clock, rise handsomely. But on all other occasions it was rarely that we could entice them up through the twelve or fifteen feet of water. Earlier in the season they are not so lazy and indifferent, but the August languor and drowsiness were now upon them. So we learned by a lucky accident to fish deep for them, even weighting our leaders with a shot, and allowing the flies to sink nearly to the bottom. After a moment's pause we would draw them slowly up, and when half or two thirds of the way to the top the trout would strike, when the sport became lively enough. Most of our fish were taken in this way. There is nothing like the flash and the strike at the surface, and perhaps only the need of food will ever tempt the genuine angler into any more prosaic style of fishing; but if you must go below the surface, a shotted leader is the best thing to use.

Our camp-fire at night served more purposes than one; from its embers and flickering shadows, Uncle Nathan read us many a tale of his life in the woods. They were the same old hunter's stories, except that they evidently had the merit of being strictly true, and hence were not very thrilling or marvelous. Uncle Nathan's tendency was rather to tone down and belittle his experiences than to exaggerate them. If he ever bragged at all (and I suspect he did just a little, when telling us how he outshot one of the famous riflemen of the American team, whom he was guiding through these woods), he did it in such a sly, round-about way that it was hard to catch him at it. His passage with the rifleman referred to shows the difference between the practical off-hand skill of the hunter in the woods and the science of the long-range target hitter. Mr. Bull's Eye had heard that his guide was a capital shot and had seen some proof of it, and hence could not rest till he had had a trial of skill with him. Uncle Nathan, being the challenged party, had the right to name the distance and the conditions. A piece of white paper the size of a silver dollar was put upon a tree twelve rods off, the contestants to fire three shots each off-hand. Uncle Nathan's first bullet barely missed the mark, but the other two were planted well into it. Then the great rifleman took his turn, and missed every time.

"By hemp!" said Uncle Nathan, "I was sorry I shot so well, Mr. ——— took it so to heart; and I had used his own rifle, too. He did not get over it for a week."

But far more ignominious was the failure of Mr. Bull's Eye when he saw his first bear. They were paddling slowly and silently down Dead River, when the guide heard a slight noise in the bushes just behind a little bend. He whispered to the rifleman, who sat kneeling in the bow of the boat, to take his rifle. But instead of doing so he picked up his two-barreled shot-gun. As they turned the point, there stood a bear not twenty yards away, drinking from the stream. Uncle Nathan held the canoe, while the man who had come so far in quest of this very game was trying to lay down his shot-gun and pick up his rifle. "His hand moved like the hand of a clock," said Uncle Nathan, "and I could hardly keep my seat. I knew the bear would see us in a moment more, and run." Instead of laying his gun by his side, where it belonged, he reached it across in front of him and laid it upon his rifle, and in trying to get the latter from under it a noise was made; the bear heard it and raised his head. Still there was time, for as the bear sprang into the woods he stopped and looked back,—"as I knew he would," said the guide; yet the marksman was not ready. "By hemp! I could have shot three bears," exclaimed Uncle Nathan, "while he was getting that rifle to his face!"

Poor Mr. Bull's Eye was deeply humiliated. "Just the chance I had been looking for," he said, "and my wits suddenly left me."

As a hunter Uncle Nathan always took the game on its own terms, that of still-hunting. He even shot foxes in this way, going into the fields in the fall just at break of day, and watching for them about their mousing haunts. One morning, by these tactics, he shot a black fox; a fine specimen, he said, and a wild one, for he stopped and looked and listened every few yards.

He had killed over two hundred moose, a large number of them at night on the lakes. His method was to go out in his canoe and conceal himself by some point or island, and wait till he heard the game. In the fall the moose comes into the water to eat the large fibrous roots of the pond-lilies. He splashes along till he finds a suitable spot, when he begins feeding, sometimes thrusting his bead and neck several feet under water. The hunter listens, and when the moose lifts his head and the rills of water run from it, and he hears him "swash" the lily roots about to get off the mud, it is his time to start. Silently as a shadow he creeps up on the moose, who by the way, it seems, never expects the approach of danger from the water side. If the hunter accidentally makes a noise the moose looks toward the shore for it. There is always a slight gleam on the water, Uncle Nathan says, even in the darkest night, and the dusky form of the moose can be distinctly seen upon it. When the hunter sees this darker shadow he lifts his gun to the sky and gets the range of its barrels, then lowers it till it covers the mark, and fires.

The largest moose Uncle Nathan ever killed is mounted in the State House at Augusta. He shot him while hunting in winter on snow-shoes. The moose was reposing upon the ground, with his head stretched out in front of him, as one may sometimes see a cow resting. The position was such that only a quartering shot through the animal's hip could reach its heart. Studying the problem carefully, and taking his own time, the hunter fired. The moose sprang into the air, turned, and came with tremendous strides straight toward him. "I knew he had not seen or scented me," said Uncle Nathan, "but, by hemp, I wished myself somewhere else just then; for I was lying right down in his path." But the noble animal stopped, a few yards short, and fell dead with a bullet-hole through his heart.

When the moose yard in the winter, that is, restrict their wanderings to a well-defined section of the forest or mountain, trampling down the snow and beating paths in all directions, they browse off only the most dainty morsels first; when they go over the ground a second time they crop a little cleaner; the third time they sort still closer, till by and by nothing is left. Spruce, hemlock, poplar, the barks of various trees, everything within reach, is cropped close. When the hunter comes upon one of these yards the problem for him to settle is, Where are the moose? for it is absolutely necessary that he keep on the lee side of them. So he considers the lay of the land, the direction of the wind, the time of day, the depth of the snow, examines the spoor, the cropped twigs, and studies every hint and clew like a detective. Uncle Nathan said he could not explain to another how he did it, but he could usually tell in a few minutes in what direction to look for the game. His experience had ripened into a kind of intuition or winged reasoning that was above rules.

He said that most large game, deer, caribou, moose, bear, when started by the hunter and not much scared, were sure to stop and look back before disappearing from sight: he usually waited for this last and best chance to fire. He told us of a huge bear he had seen one morning while still-hunting foxes in the fields; the bear saw him, and got into the woods before he could get a good shot. In her course some distance up the mountain was a bald, open spot, and he felt sure when she crossed this spot she would pause and look behind her; and sure enough, like Lot's wife, her curiosity got the better of her; she stopped to have a final look, and her travels ended there and then.

Uncle Nathan had trapped and shot a great many bears, and some of his experiences revealed an unusual degree of sagacity in this animal. One April, when the weather began to get warm and thawy, an old bear left her den in the rocks and built a large, warm nest of grass, leaves, and the bark of the white cedar, under a tall balsam fir that stood in a low, sunny, open place amid the mountains. Hither she conducted her two cubs, and the family began life in what might be called their spring residence. The tree above them was for shelter, and for refuge for the cubs in case danger approached, as it soon did in the form of Uncle Nathan. He happened that way soon after the bear had moved. Seeing her track in the snow, he concluded to follow it. When the bear had passed, the snow had been soft and sposhy, and she had "slumped," he said, several inches. It was now hard and slippery. As he neared the tree the track turned and doubled, and tacked this way and that, and led through the worst brush and brambles to be found. This was a shrewd thought of the old bear; she could thus hear her enemy coming a long time before he drew very near. When Uncle Nathan finally reached the nest, he found it empty, but still warm. Then he began to circle about and look for the bear's footprints or nail-prints upon the frozen snow. Not finding them the first time, he took a larger circle, then a still larger; finally he made a long detour, and spent nearly an hour searching for some clew to the direction the bear had taken, but all to no purpose. Then he returned to the tree and scrutinized it. The foliage was very dense, but presently he made out one of the cubs near the top, standing up amid the branches, and peering down at him. This he killed. Further search only revealed a mass of foliage apparently more dense than usual, but a bullet sent into it was followed by loud whimpering and crying, and the other baby bear came tumbling down. In leaving the place, greatly puzzled as to what had become of the mother bear, Uncle Nathan followed another of her frozen tracks, and after about a quarter of a mile saw beside it, upon the snow, the fresh trail he had been in search of. In making her escape the bear had stepped exactly in her old tracks that were hard and icy, and had thus left no mark till she took to the snow again.

During his trapping expeditions into the woods in midwinter, I was curious to know how Uncle Nathan passed the nights, as we were twice pinched with the cold at that season in our tent and blankets. It was no trouble to keep warm, he said, in the coldest weather. As night approached, he would select a place for his camp on the side of a hill. With one of his snow-shoes he would shovel out the snow till the ground was reached, carrying the snow out in front, as we scrape the earth out of the side of a hill to level up a place for the house and yard. On this level place, which, however, was made to incline slightly toward the hill, his bed of boughs was made. On the ground he had uncovered he built his fire. His bed was thus on a level with the fire, and the heat could not thaw the snow under him and let him down, or the burning logs roll upon him. With a steep ascent behind it the fire burned better, and the wind was not so apt to drive the smoke and blaze in upon him. Then, with the long, curving branches of the spruce stuck thickly around three sides of the bed, and curving over and uniting their tops above it, a shelter was formed that would keep out the cold and the snow, and that would catch and retain the warmth of the fire. Rolled in his blanket in such a nest, Uncle Nathan had passed hundreds of the most frigid winter nights.

One day we made an excursion of three miles through the woods to Bald Mountain, following a dim trail. We saw, as we filed silently along, plenty of signs of caribou, deer, and bear, but were not blessed with a sight of either of the animals themselves. I noticed that Uncle Nathan, in looking through the woods, did not hold his head as we did, but thrust it slightly forward, and peered under the branches like a deer or other wild creature.

The summit of Bald Mountain was the most impressive mountain-top I had ever seen, mainly, perhaps, because it was one enormous crown of nearly naked granite. The rock had that gray, elemental, eternal look which granite alone has. One seemed to be face to face with the gods of the fore-world. Like an atom, like a breath of to-day, we were suddenly confronted by abysmal geologic time,—the eternities past and the eternities to come. The enormous cleavage of the rocks, the appalling cracks and fissures, the rent boulders, the smitten granite floors, gave one a new sense of the power of heat and frost. In one place we noticed several deep parallel grooves, made by the old glaciers. In the depressions on the summit there was a hard, black, peaty-like soil that looked indescribably ancient and unfamiliar. Out of this mould, that might have come from the moon or the interplanetary spaces, were growing mountain cranberries and blueberries or huckleberries. We were soon so absorbed in gathering the latter that we were quite oblivious of the grandeurs about us. It is these blueberries that attract the bears. In eating them, Uncle Nathan said, they take the bushes in their mouths, and by an upward movement strip them clean of both leaves and berries. We were constantly on the lookout for the bears, but failed to see any. Yet a few days afterward, when two of our party returned here and encamped upon the mountain, they saw five during their stay, but failed to get a good shot. The rifle was in the wrong place each time. The man with the shot-gun saw an old bear and two cubs lift themselves from behind a rock and twist their noses around for his scent, and then shrink away. They were too far off for his buckshot. I must not forget the superb view that lay before us, a wilderness of woods and waters stretching away to the horizon on every band. Nearly a dozen lakes and ponds could be seen, and in a clearer atmosphere the foot of Moosehead Lake would have been visible. The highest and most striking mountain to be seen was Mount Bigelow, rising above Dead River, far to the west, and its two sharp peaks notching the horizon like enormous saw-teeth. We walked around and viewed curiously a huge boulder on the top of the mountain that had been split in two vertically, and one of the halves moved a few feet out of its bed. It looked recent and familiar, but suggested gods instead of men. The force that moved the rock had plainly come from the north. I thought of a similar boulder I had seen not long before on the highest point of the Shawangunk Mountains in New York, one side of which is propped up with a large stone, as wall-builders prop up a rock to wrap a chain around it. The rock seems poised lightly, and has but a few points of bearing. In this instance, too, the power had come from the north.

The prettiest botanical specimen my trip yielded was a little plant that bears the ugly name of horned bladderwort (Utricularia cornuta), and which I found growing in marshy places along the shores of Moxie Lake. It has a slender, naked stem nearly a foot high, crowned by two or more large deep yellow flowers,—flowers the shape of little bonnets or hoods. One almost expected to see tiny faces looking out of them. This illusion is heightened by the horn or spur of the flower, which projects from the hood like a long tapering chin,—some masker's device. Then the cape behind,—what a smart upward curve it has, as if spurned by the fairy shoulders it was meant to cover! But perhaps the most notable thing about the flower was its fragrance,—the richest and strongest perfume I have ever found in a wild flower. This our botanist, Gray, does not mention; as if one should describe the lark and forget its song. The fragrance suggested that of white clover, but was more rank and spicy.

The woods about Moxie Lake were literally carpeted with

Linnæa. I had never seen it in such profusion. In early summer, the period of its bloom, what a charming spectacle the mossy floors of these remote woods must present! The flowers are purple rose-color, nodding and fragrant. Another very abundant plant in these woods was the Clintonia borealis. Uncle Nathan said it was called "bear's corn," though he did not know why. The only noticeable flower by the Maine roadsides at this season that is not common in other parts of the country is the harebell. Its bright blue, bell-shaped corolla shone out from amid the dry grass and weeds all along the route. It was one of the most delicate roadside flowers I had ever seen.

The only new bird I saw in Maine was the pileated woodpecker, or black "log cock," called by Uncle Nathan "wood cock." I had never before seen or heard this bird, and its loud cackle in the woods about Moxie was a new sound to me. It is the wildest and largest of our northern woodpeckers, and the rarest. Its voice and the sound of its hammer are heard only in the depths of the northern woods. It is about as large as a crow, and nearly as black.

We stayed a week at Moxie, or until we became surfeited with its trout, and had killed the last Merganser duck that lingered about our end of the lake. The trout that had accumulated on our hands we had kept alive in a large champagne basket submerged in the lake, and the morning we broke camp the basket was towed to the shore and opened; and after we had feasted our eyes upon the superb spectacle, every trout, twelve or fifteen in number, some of them two-pounders, was allowed to swim back into the lake. They went leisurely, in couples and in trios, and were soon kicking up their heels in their old haunts. I expect that the divinity who presides over Moxie will see to it that every one of those trout, doubled in weight, comes to our basket in the future.

WINTER NEIGHBORS.

The country is more of a wilderness, more of a wild solitude, in the winter than in the summer. The wild comes out. The urban, the cultivated, is hidden or negatived. You shall hardly know a good field from a poor, a meadow from a pasture, a park from a forest. Lines and boundaries are disregarded; gates and bar-ways are unclosed; man lets go his hold upon the earth; title-deeds are deep buried beneath the snow; the best-kept grounds relapse to a state of nature; under the pressure of the cold all the wild creatures become outlaws, and roam abroad beyond their usual haunts. The partridge comes to the orchard for buds; the rabbit comes to the garden and lawn; the crows and jays come to the ash-heap and corn-crib, the snow-buntings to the stack and to the barn-yard; the sparrows pilfer from the domestic fowls; the pine grosbeak comes down from the north and shears your maples of their buds; the fox prowls about your premises at night, and the red squirrels find your grain in the barn or steal the butternuts from your attic. In fact, winter, like some great calamity, changes the status of most creatures and sets them adrift. Winter, like poverty, makes us acquainted with strange bedfellows.

For my part, my nearest approach to a strange bedfellow is the little gray rabbit that has taken up her abode under my study floor. As she spends the day here and is out larking at night, she is not much of a bedfellow after all. It is probable that I disturb her slumbers more than she does mine. I think she is some support to me under there-a silent wild-eyed witness and backer; a type of the gentle and harmless in savage nature. She has no sagacity to give me or lend me, but that soft, nimble foot of hers, and that touch as of cotton wherever she goes, are worthy of emulation. I think I can feel her good-will through the floor, and I hope she can mine. When I have a happy thought I imagine her ears twitch, especially when I think of the sweet apple I will place by her doorway at night. I wonder if that fox chanced to catch a glimpse of her the other night when he stealthily leaped over the fence near by and walked along between the study and the house? How clearly one could read that it was not a little dog that had passed there. There was something furtive in the track; it shied off away from the house and around it, as if eying it suspiciously; and then it had the caution and deliberation of the fox—bold, bold, but not too bold; wariness was in every footprint. If it had been a little dog that had chanced to wander that way, when he crossed my path he would have followed it up to the barn and have gone smelling around for a bone; but this sharp, cautious track held straight across all others, keeping five or six rods from the house, up the hill, across the highway towards a neighboring farmstead, with its nose in the air and its eye and ear alert, so to speak.

A winter neighbor of mine in whom I am interested, and who perhaps lends me his support after his kind, is a little red owl, whose retreat is in the heart of an old apple-tree just over the fence. Where he keeps himself in spring and summer I do not know, but late every fall, and at intervals all winter, his hiding-place is discovered by the jays and nut-hatches, and proclaimed from the tree-tops for the space of half an hour or so, with all the powers of voice they can command. Four times during one winter they called me out to behold this little ogre feigning sleep in his den, sometimes in one apple-tree, sometimes in another. Whenever I heard their cries, I knew my neighbor was being berated. The birds would take turns at looking in upon him and uttering their alarm-notes. Every jay within hearing would come to the spot and at once approach the hole in the trunk or limb, and with a kind of breathless eagerness and excitement take a peep at the owl, and then join the outcry. When I approached they would hastily take a final look and then withdraw and regard my movements intently. After accustoming my eye to the faint light of the cavity for a few moments, I could usually make out the owl at the bottom feigning sleep. Feigning, I say, because this is what he really did, as I first discovered one day when I cut into his retreat with the axe. The loud blows and the falling chips did not disturb him at all. When I reached in a stick and pulled him over on his side, leaving one of his wings spread out, he made no attempt to recover himself, but lay among the chips and fragments of decayed wood, like a part of themselves. Indeed, it took a sharp eye to distinguish him. Nor till I had pulled him forth by one wing, rather rudely, did he abandon his trick of simulated sleep or death. Then, like a detected pickpocket, he was suddenly transformed into another creature. His eyes flew wide open, his talons clutched my finger, his ears were depressed, and every motion and look said, "Hands off, at your peril." Finding this game did not work, he soon began to "play 'possum" again. I put a cover over my study wood-box and kept him captive for a week. Look in upon him any time, night or day, and he was apparently wrapped in the profoundest slumber; but the live mice which I put into his box from time to time found his sleep was easily broken; there would be a sudden rustle in the box, a faint squeak, and then silence. After a week of captivity I gave him his freedom in the full sunshine: no trouble for him to see which way and where to go.

Just at dusk in the winter nights, I often hear his soft bur-r-r-r, very pleasing and bell-like. What a furtive, woody sound it is in the winter stillness, so unlike the harsh scream of the hawk. But all the ways of the owl are ways of softness and duskiness. His wings are shod with silence, his plumage is edged with down.

Another owl neighbor of mine, with whom I pass the time of day more frequently than with the last, lives farther away. I pass his castle every night on my way to the post-office, and in winter, if the hour is late enough, am pretty sure to see him standing in his doorway, surveying the passers-by and the landscape through narrow slits in his eyes. For four successive winters now have I observed him. As the twilight begins to deepen he rises out of his cavity in the apple-tree, scarcely faster than the moon rises from behind the hill, and sits in the opening, completely framed by its outlines of gray bark and dead wood, and by his protective coloring virtually invisible to every eye that does not know he is there. Probably my own is the only eye that has ever penetrated his secret, and mine never would have done so had I not chanced on one occasion to see him leave his retreat and make a raid upon a shrike that was impaling a shrew-mouse upon a thorn in a neighboring tree and which I was watching. Failing to get the mouse, the owl returned swiftly to his cavity, and ever since, while going that way, I have been on the lookout for him. Dozens of teams and foot-passengers pass him late in the day, but he regards them not, nor they him. When I come alone and pause to salute him, he opens his eyes a little wider, and, appearing to recognize me, quickly shrinks and fades into the background of his door in a very weird and curious manner. When he is not at his outlook, or when he is, it requires the best powers of the eye to decide the point, as the empty cavity itself is almost an exact image of him. If the whole thing had been carefully studied it could not have answered its purpose better. The owl stands quite perpendicular, presenting a front of light mottled gray; the eyes are closed to a mere slit, the ear-feathers depressed, the beak buried in the plumage, and the whole attitude is one of silent, motionless waiting and observation. If a mouse should be seen crossing the highway, or scudding over any exposed part of the snowy surface in the twilight, the owl would doubtless swoop down upon it. I think the owl has learned to distinguish me from the rest of the passers-by; at least, when I stop before him, and he sees himself observed, he backs down into his den, as I have said, in a very amusing manner. Whether bluebirds, nut-hatches, and chickadees—birds that pass the night in cavities of trees—ever run into the clutches of the dozing owl, I should be glad to know. My impression is, however, that they seek out smaller cavities. An old willow by the roadside blew down one summer, and a decayed branch broke open, revealing a brood of half-fledged owls, and many feathers and quills of bluebirds, orioles, and other songsters, showing plainly enough why all birds fear and berate the owl.

The English house sparrows, that are so rapidly increasing among us, and that must add greatly to the food supply of the owls and other birds of prey, seek to baffle their enemies by roosting in the densest evergreens they can find, in the arbor-vitæ, and in hemlock hedges. Soft-winged as the owl is, he cannot steal in upon such a retreat without giving them warning.

These sparrows are becoming about the most noticeable of my winter neighbors, and a troop of them every morning watch me put out the hens' feed, and soon claim their share. I rather encouraged them in their neighborliness, till one day I discovered the snow under a favorite plum-tree where they most frequently perched covered with the scales of the fruit-buds. On investigating I found that the tree had been nearly stripped of its buds—a very unneighborly act on the part of the sparrows, considering, too, all the cracked corn I had scattered for them. So I at once served notice on them that our good understanding was at an end. And a hint is as good as a kick with this bird. The stone I hurled among them, and the one with which I followed them up, may have been taken as a kick; but they were only a hint of the shot-gun that stood ready in the corner. The sparrows left in high dungeon, and were not back again in some days, and were then very shy. No doubt the time is near at hand when we shall have to wage serious war upon these sparrows, as they long have had to do on the continent of Europe. And yet it will be hard to kill the little wretches, the only Old World bird we have. When I take down my gun to shoot them I shall probably remember that the Psalmist said, "I watch, and am as a sparrow alone upon the house-top," and maybe the recollection will cause me to stay my hand. The sparrows have the Old World hardiness and prolificness; they are wise and tenacious of life, and we shall find it by and by no small matter to keep them in check. Our native birds are much different, less prolific, less shrewd, less aggressive and persistent, less quick-witted and able to read the note of danger or hostility—in short, less sophisticated. Most of our birds are yet essentially wild, that is, little changed by civilization. In winter, especially, they sweep by me and around me in flocks,—the Canada sparrow, the snow-bunting, the shore-lark, the pine grosbeak, the red-poll, the cedar-bird,— feeding upon frozen apples in the orchard, upon cedar-berries, upon maple-buds, and the berries of the mountain ash, and the celtis, and upon the seeds of the weeds that rise above the snow in the field, or upon the hay-seed dropped where the cattle have been foddered in the barn-yard or about the distant stack; but yet taking no heed of man, in no way changing their habits so as to take advantage of his presence in nature. The pine grosbeak will come in numbers upon your porch, to get the black drupes of the honeysuckle or the woodbine, or within reach of your windows to get the berries of the mountain-ash, but they know you not; they look at you as innocently and unconcernedly as at a bear or moose in their native north, and your house is no more to them than a ledge of rocks.

The only ones of my winter neighbors that actually rap at my door are the nut-hatches and woodpeckers, and these do not know that it is my door. My retreat is covered with the bark of young chestnut-trees, and the birds, I suspect, mistake it for a huge stump that ought to hold fat grubs (there is not even a bookworm inside of it), and their loud rapping often makes me think I have a caller indeed. I place fragments of hickory-nuts in the interstices of the bark, and thus attract the nut-hatches; a bone upon my window-sill attracts both nut-hatches and the downy woodpecker. They peep in curiously through the window upon me, pecking away at my bone, too often a very poor one. A bone nailed to a tree a few feet in front of the window attracts crows as well as lesser birds. Even the slate-colored snow-bird, a seed-eater, comes and nibbles it occasionally.

The bird that seems to consider he has the best right to the bone both upon the tree and upon the sill is the downy woodpecker, my favorite neighbor among the winter birds, to whom I will mainly devote the remainder of this chapter. His retreat is but a few paces from my own, in the decayed limb of an apple-tree which he excavated several autumns ago. I say "he" because the red plume on the top of his head proclaims the sex. It seems not to be generally known to our writers upon ornithology that certain of our woodpeckers—probably all the winter residents—each fall excavate a limb or the trunk of a tree in which to pass the winter, and that the cavity is abandoned in the spring, probably for a new one in which nidification takes place. So far as I have observed, these cavities are drilled out only by the males. Where the females take up their quarters I am not so well informed, though I suspect that they use the abandoned holes of the males of the previous year.

The particular woodpecker to which I refer drilled his first hole in my apple-tree one fall four or five years ago. This he occupied till the following spring when he abandoned it. The next fall he began a hole in an adjoining limb, later than before, and when it was about half completed a female took possession of his old quarters. I am sorry to say that this seemed to enrage the male, very much, and he persecuted the poor bird whenever she appeared upon the scene. He would fly at her spitefully and drive her off. One chilly November morning, as I passed under the tree, I heard the hammer of the little architect in his cavity, and at the same time saw the persecuted female sitting at the entrance of the other hole as if she would fain come out. She was actually shivering, probably from both fear and cold. I understood the situation at a glance; the bird was afraid to come forth and brave the anger of the male. Not till I had rapped smartly upon the limb with my stick did she come out and attempt to escape; but she had not gone ten feet from the tree before the male was in hot pursuit, and in a few moments had driven her back to the same tree, where she tried to avoid him among the branches. A few days after, he rid himself of his unwelcome neighbor in the following ingenious manner: he fairly scuttled the other cavity; he drilled a hole into the bottom of it that let in the light and the cold, and I saw the female there no more. I did not see him in the act of rendering this tenement uninhabitable; but one morning, behold it was punctured at the bottom, and the circumstances all seemed to point to him as the author of it. There is probably no gallantry among the birds except at the mating season. I have frequently seen the male woodpecker drive the female away from the bone upon the tree. When she hopped around to the other end and timidly nibbled it, he would presently dart spitefully at her. She would then take up her position in his rear and wait till he had finished his meal. The position of the female among the birds is very much the same as that of woman among savage tribes. Most of the drudgery of life falls upon her, and the leavings of the males are often her lot.

My bird is a genuine little savage, doubtless, but I value him as a neighbor. It is a satisfaction during the cold or stormy winter nights to know he is warm and cosy there in his retreat. When the day is bad and unfit to be abroad in; he is there too. When I wish to know if he is at home, I go and rap upon his tree, and, if he is not too lazy or indifferent, after some delay he shows his head in his round doorway about ten feet above, and looks down inquiringly upon me—sometimes latterly I think half resentfully, as much as to say, "I would thank you not to disturb me so often." After sundown, he will not put his head out any more when I call, but as I step away I can get a glimpse of him inside looking cold and reserved. He is a late riser, especially if it is a cold or disagreeable morning, in this respect being like the fowls; it is sometimes near nine o'clock before I see him leave his tree. On the other hand, he comes home early, being in if the day is unpleasant by four P. M. He lives all alone; in this respect I do not commend his example. Where his mate is I should like to know.

I have discovered several other woodpeckers in adjoining orchards, each of which has a like home and leads a like solitary life. One of them has excavated a dry limb within easy reach of my hand, doing the work also in September. But the choice of tree was not a good one; the limb was too much decayed, and the workman had made the cavity too large; a chip had come out, making a hole in the outer wall. Then he went a few inches down the limb and began again, and excavated a large, commodious chamber, but had again come too near the surface; scarcely more than the bark protected him in one place, and the limb was very much weakened. Then he made another attempt still farther down the limb, and drilled in an inch or two, but seemed to change his mind; the work stopped, and I concluded the bird had wisely abandoned the tree. Passing there one cold, rainy November day, I thrust in my two fingers and was surprised to feel something soft and warm: as I drew away my hand the bird came out, apparently no more surprised than I was. It had decided, then, to make its home in the old limb; a decision it had occasion to regret, for not long after, on a stormy night, the branch gave way and fell to the ground.

> *"When the bough breaks the cradle will fall,*
> *and down will come baby, cradle and all."*

Such a cavity makes a snug, warm home, and when the entrance is on the under side if the limb, as is usual, the wind and snow cannot reach the occupant. Late in December, while crossing a high, wooded mountain, lured by the music of fox-hounds, I discovered fresh yellow chips strewing the new-fallen snow, and at once thought of my woodpeckers. On looking around I saw where one had been at work excavating a lodge in a small yellow birch. The orifice was about fifteen feet from the ground, and appeared as round as if struck with a compass. It was on the east side of the tree, so as to avoid the prevailing west and northeast winds. As it was nearly two inches in diameter, it could not have been the work of the downy, but must have been that of the hairy, or else the yellow-bellied woodpecker. His home had probably been wrecked by some violent wind, and he was thus providing himself another. In digging out these retreats the woodpeckers prefer a dry, brittle, trunk, not too soft. They go in horizontally to the centre and then turn downward, enlarging the tunnel as they go, till when finished it is the shape of a long, deep pear.

Another trait our woodpeckers have that endears them to me, and that has never been pointedly noticed by our ornithologists, is their habit of drumming in the spring. They are songless birds, and yet all are musicians; they make the dry limbs eloquent of the coming change. Did you think that loud, sonorous hammering which proceeded from the orchard or from the near woods on that still March or April morning was only some bird getting its breakfast? It is downy, but he is not rapping at the door of a grub; he is rapping at the door of spring, and the dry limb thrills beneath the ardor of his blows. Or, later in the season, in the dense forest or by some remote mountain lake, does that measured rhythmic beat that breaks upon the silence, first three strokes following each other rapidly, succeeded by two louder ones with longer intervals between them, and that has an effect upon the alert ear as if the solitude itself had at last found a voice—does that suggest anything less than a deliberate musical performance? In fact, our woodpeckers are just as characteristically drummers as is the ruffed grouse, and they have their particular limbs and stubs to which they resort for that purpose. Their need of expression is apparently just as great as that of the song-birds, and it is not surprising that they should have found out that there is music in a dry, seasoned limb which can be evoked beneath their beaks.

A few seasons ago a downy woodpecker, probably the individual one who is now my winter neighbor, began to drum early in March in a partly decayed apple-tree that stands in the edge of a narrow strip of woodland near me. When the morning was still and mild I would often hear him through my window before I was up, or by half-past six o'clock, and he would keep it up pretty briskly till nine or ten o'clock, in this respect resembling the grouse, which do most of their drumming in the forenoon. His drum was the stub of a dry limb about the size of one's wrist.

The heart was decayed and gone, but the outer shell was hard and resonant. The bird would keep his position there for an hour at a time. Between his drummings he would preen his plumage and listen as if for the response of the female, or for the drum of some rival. How swift his head would go when he was delivering his blows upon the limb! His beak wore the surface perceptibly. When he wished to change the key, which was quite often, he would shift his position an inch or two to a knot which gave out a higher, shriller note. When I climbed up to examine his drum he was much disturbed. I did not know he was in the vicinity, but it seems he saw me from a near tree, and came in haste to the neighboring branches, and with spread plumage and a sharp note demanded plainly enough what my business was with his drum. I was invading his privacy, desecrating his shrine, and the bird was much put out. After some weeks the female appeared; he had literally drummed up a mate; his urgent and oft-repeated advertisement was answered. Still the drumming did not cease, but was quite as fervent as before. If a mate could be won by drumming she could be kept and entertained by more drumming; courtship should not end with marriage. If the bird felt musical before, of course he felt much more so now. Besides that, the gentle deities needed propitiating in behalf of the nest and young as well as in behalf of the mate. After a time a second female came, when there was war between the two. I did not see them come to blows, but I saw one female pursuing the other about the place, and giving her no rest for several days. She was evidently trying to run her out of the neighborhood. Now and then she, too, would drum briefly as if sending a triumphant message to her mate.

The woodpeckers do not each have a particular dry limb to which they resort at all times to drum, like the one I have described. The woods are full of suitable branches, and they drum more or less here and there as they are in quest of food; yet I am convinced each one has its favorite spot, like the grouse, to which it resorts, especially in the morning. The sugar-maker in the maple-woods may notice that their sound proceeds from the same tree or trees about his camp with great regularity. A woodpecker in my vicinity has drummed for two seasons on a telegraph pole, and he makes the wires and glass insulators ring. Another drums on a thin board on the end of a long grape-arbor, and on still mornings can be heard a long distance.

A friend of mine in a Southern city tells me of a red-headed woodpecker that drums upon a lightning-rod on his neighbor's house. Nearly every clear, still morning at certain seasons, he says, this musical rapping may be heard. "He alternates his tapping with his stridulous call, and the effect on a cool, autumn-like morning is very pleasing."

The high-hole appears to drum more promiscuously than does the downy. He utters his long, loud spring call, whick—whick—whick—whick, and then begins to rap with his beak upon his perch before the last note has reached your ear. I have seen him drum sitting upon the ridge of the barn. The log cock, or pileated woodpecker, the largest and wildest of our Northern species, I have never heard drum. His blows should wake the echoes.

When the woodpecker is searching for food, or laying siege to some hidden grub, the sound of his hammer is dead or muffled, and is heard but a few yards. It is only upon dry, seasoned timber, freed of its bark, that he beats his reveille to spring and wooes his mate.

Wilson was evidently familiar with this vernal drumming of the woodpeckers, but quite misinterprets it. Speaking of the red-bellied species, he says: "It rattles like the rest of the tribe on the dead limbs, and with such violence as to be heard in still weather more than half a mile off; and listens to hear the insect it has alarmed." He listens rather to hear the drum of his rival or the brief and coy response of the female; for there are no insects in these dry limbs.

On one occasion I saw downy at his drum when a female flew quickly through the tree and alighted a few yards beyond him. He paused instantly, and kept his place, apparently without moving a muscle. The female, I took it, had answered his advertisement. She flitted about from limb to limb (the female may be known by the absence of the crimson spot on the back of the head), apparently full of business of her own, and now and then would drum in a shy, tentative manner. The male watched her a few moments and, convinced perhaps that she meant business, struck up his liveliest tune, then listened for her response. As it came back timidly but promptly, he left his perch and sought a nearer acquaintance with the prudent female. Whether or not a match grew out of this little flirtation I cannot say.

Our smaller woodpeckers are sometimes accused of injuring the apple and other fruit trees, but the depredator is probably the larger and rarer yellow-bellied species. One autumn I caught one of these fellows in the act of sinking long rows of his little wells in the limb of an apple-tree. There were series of rings of them, one above another, quite around the stem, some of them the third of an inch across. They are evidently made to get at the tender, juicy bark, or cambium layer, next to the hard wood of the tree. The health and vitality of the branch are so seriously impaired by them that it often dies.

In the following winter the same bird (probably) tapped a maple-tree in front of my window in fifty six places; and when the day was sunny, and the sap oozed out, he spent most of his time there. He knew the good sap-days, and was on hand promptly for his tipple; cold and cloudy days he did not appear. He knew which side of the tree to tap, too, and avoided the sunless northern exposure. When one series of well-holes failed to supply him, he would sink another, drilling through the bark with great ease and quickness. Then, when the day was warm, and the sap ran freely, he would have a regular sugar-maple debauch, sitting there by his wells hour after hour, and as fast as they became filled sipping out the sap. This he did in a gentle, caressing manner that was very suggestive. He made a row of wells near the foot of the tree, and other rows higher up, and he would hop up and down the trunk as these became filled. He would hop down the tree backward with the utmost ease, throwing his tail outward and his head inward at each hop. When the wells would freeze or his thirst become slaked, he would ruffle his feathers, draw himself together, and sit and doze in the sun on the side of the tree. He passed the night in a hole in an apple-tree not far off. He was evidently a young bird not yet having the plumage of the mature male or female, and yet he knew which tree to tap and where to tap it. I saw where he had bored several maples in the vicinity, but no oaks or chestnuts. I nailed up a fat bone near his sap-works: the downy woodpecker came there several times a day to dine; the nut-hatch came, and even the snow-bird took a taste occasionally; but this sap-sucker never touched it; the sweet of the tree sufficed for him. This woodpecker does not breed or abound in my vicinity; only stray specimens are now and then to be met with in the colder months. As spring approached, the one I refer to took his departure.

I must bring my account of my neighbor in the tree down to the latest date; so after the lapse of a year I add the following notes. The last day of February was bright and springlike. I heard the first sparrow sing that morning and the first screaming of the circling hawks, and about seven o'clock the first drumming of my little friend. His first notes were uncertain and at long intervals, but by and by he warmed up and beat a lively tattoo. As the season advanced he ceased to lodge in his old quarters. I would rap and find nobody at home. Was he out on a lark, I said, the spring fever working in his blood? After a time his drumming grew less frequent, and finally, in the middle of April, ceased entirely. Had some accident befallen him, or had he wandered away to fresh fields, following some siren of his species? Probably the latter. Another bird that I had under observation also left his winter-quarters in the spring. This, then, appears to be the usual custom. The wrens and the nut-hatches and chickadees succeed to these abandoned cavities, and often have amusing disputes over them. The nut-hatches frequently pass the night in them, and the wrens and chickadees nest in them. I have further observed that in excavating a cavity for a nest the downy woodpecker makes the entrance smaller than when he is excavating his winter-quarters. This is doubtless for the greater safety of the young birds.

The next fall, the downy excavated another limb in the old apple-tree, but had not got his retreat quite finished, when the large hairy woodpecker appeared upon the scene. I heard his loud click, click, early one frosty November morning. There was something impatient and angry in the tone that arrested my attention. I saw the bird fly to the tree where downy had been at work, and fall with great violence upon the entrance to his cavity. The bark and the chips flew beneath his vigorous blows, and before I fairly woke up to what he was doing, he had completely demolished the neat, round doorway of downy. He had made a large ragged opening large enough for himself to enter. I drove him away and my favorite came back, but only to survey the ruins of his castle for a moment and then go away. He lingered about for a day or two and then disappeared. The big hairy usurper passed a night in the cavity, but on being hustled out of it the next night by me, he also left, but not till he had demolished the entrance to a cavity in a neighboring tree where downy and his mate had reared their brood that summer, and where I had hoped the female would pass the winter.

NOTES BY THE WAY.

I. THE WEATHER-WISE MUSKRAT

I am more than half persuaded that the muskrat is a wise little animal, and that on the subject of the weather, especially, he possesses some secret that I should be glad to know. In the fall of 1878 I noticed that he built unusually high and massive nests. I noticed them in several different localities. In a shallow, sluggish pond by the roadside, which I used to pass daily in my walk, two nests were in process of construction throughout the month of November. The builders worked only at night, and I could see each day that the work had visibly advanced. When there was a slight skim of ice over the pond, this was broken up about the nests, with trails through it in different directions where the material had been brought. The houses were placed a little to one side of the main channel, and were constructed entirely of a species of coarse wild grass that grew all about. So far as I could see, from first to last they were solid masses of grass, as if the interior cavity or nest was to be excavated afterward, as doubtless it was. As they emerged from the pond they gradually assumed the shape of a miniature mountain, very bold and steep on the south side, and running down a long gentle grade to the surface of the water on the north. One could see that the little architect hauled all his material up this easy slope, and thrust it out boldly around the other side. Every mouthful was distinctly defined. After they were two feet or more above the water, I expected each day to see that the finishing stroke had been given and the work brought to a close. But higher yet, said the builder. December drew near, the cold became threatening, and I was apprehensive that winter would suddenly shut down upon those unfinished nests. But the wise rats knew better than I did; they had received private advices from headquarters that I knew not of. Finally, about the 6th of December, the nests assumed completion; the northern incline was absorbed or carried up, and each structure became a strong massive cone, three or four feet high, the largest nest of the kind I had ever seen. Does it mean a severe winter? I inquired. An old farmer said it meant "high water," and he was right once, at least, for in a few days afterward we had the heaviest rainfall known in this section for half a century. The creeks rose to an almost unprecedented height. The sluggish pond became a seething, turbulent watercourse; gradually the angry element crept up the sides of these lake dwellings, till, when the rain ceased, about four o'clock they showed above the flood no larger than a man's hat. During the night the channel shifted till the main current swept over them, and next day not a vestige of the nests was to be seen; they had gone down-stream, as had many other dwellings of a less temporary character. The rats had built wisely, and would have been perfectly secure against any ordinary high water, but who can foresee a flood? The oldest traditions of their race did not run back to the time of such a visitation.

Nearly a week afterward another dwelling was begun, well away from the treacherous channel, but the architects did not work at it with much heart; the material was very scarce, the ice hindered, and before the basement-story was fairly finished, winter had the pond under his lock and key.

In other localities I noticed that where the nests were placed on the banks of streams, they were made secure against the floods by being built amid a small clump of bushes. When the fall of 1879 came, the muskrats were very tardy about beginning their house, laying the corner-stone—or the corner-sod-about

December 1st, and continuing the work slowly and indifferently. On the 15th of the month the nest was not yet finished. This, I said, indicates a mild winter; and, sure enough, the season was one of the mildest known for many years. The rats had little use for their house.

Again, in the fall of 1880, while the weather-wise were wagging their heads, some forecasting a mild, some a severe winter, I watched with interest for a sign from my muskrats. About November 1st, a month earlier than the previous year, they began their nest, and worked at it with a will. They appeared to have just got tidings of what was coming. If I had taken the hint so palpably given, my celery would not have been frozen in the ground, and my apples caught in unprotected places. When the cold wave struck us, about November 20th, my four-legged "I-told-you-so's" had nearly completed their dwelling; it lacked only the ridge-board, so to speak; it needed a little "topping out," to give it a finished look. But this it never got. The winter had come to stay, and it waxed more and more severe, till the unprecedented cold of the last days of December must have astonished even the wise muskrats in their snug retreat. I approached their nest at this time, a white mound upon the white, deeply frozen surface of the pond, and wondered if there was any life in that apparent sepulchre. I thrust my walking-stick sharply into it, when there was a rustle and a splash into the water, as the occupant made his escape. What a damp basement that house has, I thought, and what a pity to rout out a peaceful neighbor out of his bed in this weather and into such a state of things as this! But water does not wet the muskrat; his fur is charmed, and not a drop penetrates it. Where the ground is favorable, the muskrats do not build these mound-like nests, but burrow into the bank a long distance, and establish their winter-quarters there.

Shall we not say, then, in view of the above facts, that this little creature is weather-wise? The hitting of the mark twice might be mere good luck; but three bull's-eyes in succession is not a mere coincidence; it is a proof of skill. The muskrat is not found in the Old World, which is a little singular, as other rats so abound there, and as those slow-going English streams especially, with their grassy banks, are so well suited to him. The water-rat of Europe is smaller, but of similar nature and habits. The muskrat does not hibernate like some rodents, but is pretty active all winter. In December I noticed in my walk where they had made excursions of a few yards to an orchard for frozen apples. One day, along a little stream, I saw a mink track amid those of the muskrat; following it up, I presently came to blood and other marks of strife upon the snow beside a stone wall. Looking in between the stones, I found the carcass of the luckless rat, with its head and neck eaten away. The mink had made a meal of him.

II. CHEATING THE SQUIRRELS.

FOR the largest and finest chestnuts I had last fall I was indebted to the gray squirrels. Walking through the early October woods one day, I came upon a place where the ground was thickly strewn with very large unopened chestnut burs. On examination I found that every bur had been cut square off with about an inch of the stem adhering, and not one had been left on the tree. It was not accident, then, but design. Whose design? The squirrels'. The fruit was the finest I had ever seen in the woods, and some wise squirrel had marked it for his own. The burs were ripe, and had just begun to divide, not "threefold," but fourfold, "to show the fruit within." The squirrel that had taken all this pains had evidently reasoned with himself thus: "Now, these are extremely fine chestnuts, and I want them; if I wait till the burs open on the tree the crows and jays will be sure to carry off a great many of the nuts before they fall; then, after the wind has rattled out what remain, there are the mice, the chipmunks, the red squirrels, the raccoons, the grouse, to say nothing of the boys and the pigs, to come in for their share; so I will forestall events a little; I will cut off the burs when they have matured, and a few days of this dry October weather will cause everyone of them to open on the ground; I shall be on hand in the nick of time to gather up my nuts." The squirrel, of course, had to take the chances of a prowler like myself coming along, but he had fairly stolen a march on his neighbors. As I proceeded to collect and open the burs, I was half prepared to hear an audible protest from the trees about, for I constantly fancied myself watched by shy but jealous eyes. It is an interesting inquiry how the squirrel knew the burs would open if left to know, but thought the experiment worth trying.

The gray squirrel is peculiarly an American product, and might serve very well as a national emblem. The Old World can beat us on rats and mice, but we are far ahead on squirrels, having five or six species to Europe's one.

III. FOX AND HOUND.

I STOOD on a high hill or ridge one autumn day and saw a hound run a fox through the fields far beneath me. What odors that fox must have shaken out of himself, I thought, to be traced thus easily, and how great their specific gravity not to have been blown away like smoke by the breeze! The fox ran a long distance down the hill, keeping within a few feet of a stone wall; then turned a right angle and led off for the mountain, across a plowed field and a succession of pasture lands. In about fifteen minutes the hound came in full blast with her nose in the air, and never once did she put it to the ground while in my sight. When she came to the stone wall she took the other side from that taken by the fox, and kept about the same distance from it, being thus separated several yards from his track, with the fence between her and it. At the point where the fox turned sharply to the left, the hound overshot a few yards, then wheeled, and feeling the air a moment with her nose, took up the scent again and was off on his trail as unerringly as fate. It seemed as if the fox must have sowed himself broadcast as he went along, and that his scent was so rank and heavy that it settled in the hollows and clung tenaciously to the bushes and crevices in the fence. I thought I ought to have caught a remnant of it as I passed that way some minutes later, but I did not. But I suppose it was not that the light-footed fox so impressed himself upon the ground he ran over, but that the sense of the hound was so keen. To her sensitive nose these tracks steamed like hot cakes, and they would not have cooled off so as to be undistinguishable for several hours. For the time being she had but one sense: her whole soul was concentrated in her nose.

It is amusing when the hunter starts out of a winter morning to see his hound probe the old tracks to determine how recent they are. He sinks his nose down deep in the snow so as to exclude the air from above, then draws a long full breath, giving sometimes an audible snort If there remains the least effluvium of the fox the hound will detect it. If it be very slight it only sets his tail wagging; if it be strong it unloosens his tongue.

Such things remind one of the waste, the friction that is going on all about us, even when the wheels of life run the most smoothly. A fox cannot trip along the top of a stone wall so lightly but that he will leave enough of himself to betray his course to the hound for hours afterward. When the boys play "hare and hounds" the hare scatters bits of paper to give a clew to the pursuers, but he scatters himself much more freely if only our sight and scent were sharp enough to detect the fragments. Even the fish leave a trail in the water, and it is said the otter will pursue them by it. The birds make a track in the air, only their enemies hunt by sight rather than by scent. The fox baffles the hound most upon a hard crust of frozen snow; the scent will not hold to the smooth, bead-like granules.

Judged by the eye alone, the fox is the lightest and most buoyant creature that runs. His soft wrapping of fur conceals the muscular play and effort that is so obvious in the hound that pursues him, and he comes bounding along precisely as if blown by a gentle wind. His massive tail is carried as if it floated upon the air by its own lightness.

The hound is not remarkable for his fleetness, but how he will hang!—often running late into the night and sometimes till morning, from ridge to ridge, from peak to peak; now on the mountain, now crossing the valley, now playing about a large slope of uplying pasture fields. At times the fox has a pretty well-defined orbit, and the hunter knows where to intercept him. Again he leads off like a comet, quite beyond the system of hills and ridges upon which he was started, and his return is entirely a matter of conjecture; but if the day be not more than half spent, the chances are that the fox will be back before night, though the sportsman's patience seldom holds out that long.

The hound is a most interesting dog. How solemn and long-visaged he is—how peaceful and well-disposed! He is the Quaker among dogs. All the viciousness and currishness seem to have been weeded out of him; he seldom quarrels, or fights, or plays, like other dogs. Two strange hounds, meeting for the first time, behave as civilly toward each other as if two men. I know a hound that has an ancient, wrinkled, human, far-away look that reminds one of the bust of Homer among the Elgin marbles. He looks like the mountains toward which his heart yearns so much.

The hound is a great puzzle to the farm dog; the latter, attracted by his baying, comes barking and snarling up through the fields bent on picking a quarrel; he intercepts the hound, snubs and insults and annoys him in every way possible, but the hound heeds him not; if the dog attacks him he gets away as best he can, and goes on with the trail; the cur bristles and barks and struts about for a while, then goes back to the house, evidently thinking the hound a lunatic, which he is for the time being—a monomaniac, the slave and victim of one idea. I saw the master of a hound one day arrest him in full course to give one of the hunters time to get to a certain runaway; the dog cried and struggled to free himself and would listen neither to threats nor caresses. Knowing he must be hungry, I offered him my lunch, but he would not touch it. I put it in his mouth, but he threw it contemptuously from him. We coaxed and petted and reassured him, but he was under a spell; he was bereft of all thought or desire but the one passion to pursue that trail.

IV. THE WOODCHUCK

Writers upon rural England and her familiar natural history make no mention of the marmot or woodchuck. In Europe this animal seems to be confined to high mountainous districts, as on our Pacific slope, burrowing near the snow line. It is more social or gregarious than the American species, living in large families like our prairie-dog. In the Middle and Eastern States our woodchuck takes the place, in some respects, of the English rabbit, burrowing in every hillside and under every stone wall and jutting ledge and large bowlder, from whence it makes raids upon the grass and clover and sometimes upon the garden vegetables. It is quite solitary in its habits, seldom more than one inhabiting the same den, unless it be a mother and her young. It is not now so much a wood chuck as a field chuck. Occasionally, however, one seems to prefer the woods, and is not seduced by the sunny slopes and the succulent grass, but feeds, as did his fathers before him, upon roots and twigs, the bark of young trees, and upon various wood plants.

One summer day, as I was swimming across a broad, deep

pool in the creek in a secluded place in the woods, I saw one of these sylvan chucks amid the rocks but a few feet from the edge of the water where I proposed to touch. He saw my approach, but doubtless took me for some water-fowl, or for some cousin of his of the muskrat tribe; for he went on with his feeding, and regarded me not till I paused within ten feet of him and lifted myself up. Then he did not know me; having, perhaps, never seen Adam in his simplicity, but he twisted his nose around to catch my scent; and the moment he had done so he sprang like a jumping-jack and rushed into his den with the utmost precipitation.

The woodchuck is the true serf among our animals; he belongs to the soil, and savors of it. He is of the earth, earthy. There is generally a decided odor about his dens and lurking-places, but it is not at all disagreeable in the clover-scented air, and his shrill whistle, as he takes to his hole or defies the farm dog from the interior of the stone wall, is a pleasant summer sound. In form and movement the woodchuck is not captivating. His body is heavy and flabby. Indeed, such a flaccid, fluid, pouchy carcass, I have never before seen. It has absolutely no muscular tension or rigidity, but is as baggy and shaky as a skin filled with water. Let the rifleman shoot one while it lies basking on a sidelong rock, and its body slumps off, and rolls and spills down the hill, as if it were a mass of bowels only. The legs of the woodchuck are short and stout, and made for digging rather than running. The latter operation he performs by short leaps, his belly scarcely clearing the ground. For a short distance he can make very good time, but he seldom trusts himself far from his hole, and when surprised in that predicament, makes little effort to escape, but, grating his teeth, looks the danger squarely in the face.

I knew a farmer in New York who had a very large bob-tailed churn-dog by the name of Cuff. The farmer kept a large dairy and made a great deal of butter, and it was the business of Cuff to spend nearly the half of each summer day treading the endless round of the churning-machine. During the remainder of the day he had plenty of time to sleep, and rest, and sit on his hips and survey the landscape. One day, sitting thus, he discovered a woodchuck about forty rods from the house, on a steep side-hill, feeding about near his hole, which was beneath a large rock. The old dog, forgetting his stiffness, and remembering the fun he had had with woodchucks in his earlier days, started off at his highest speed, vainly hoping to catch this one before he could get to his hole. But the woodchuck, seeing the dog come laboring up the hill, sprang to the mouth of his den, and, when his pursuer was only a few rods off, whistled tauntingly and went in. This occurred several times, the old dog marching up the hill, and then marching down again, having had his labor for his pains. I suspect that he revolved the subject in his mind while he revolved the great wheel of the churning-machine, and that some turn or other brought him a happy thought, for next time he showed himself a strategist. Instead of giving chase to the woodchuck when first discovered, he crouched down to the ground, and, resting his head on his paws, watched him. The woodchuck kept working away from the hole, lured by the tender clover, but, not unmindful of his safety, lifted himself up on his haunches every few moments and surveyed the approaches. Presently, after the woodchuck had let himself down from one of these attitudes of observation, and resumed his feeding, Cuff started swiftly but stealthily up the hill, precisely in the attitude of a cat when she is stalking a bird. When the woodchuck rose up again, Cuff was perfectly motionless and half hid by the grass. When he again resumed his clover, Cuff sped up the hill as before, this time crossing a fence, but in a low place, and so nimbly that he was not discovered. Again the wood chuck was on the outlook, again Cuff was motionless and hugging the ground. As the dog nears his victim he is partially hidden by a swell in the earth, but still the woodchuck from his outlook reports "all right," when Cuff, having not twice as far to run as the 'chuck, throws all stealthiness aside and rushes directly for the hole. At that moment the woodchuck discovers his danger, and, seeing that it is a race for life, leaps as I never saw marmot leap before. But he is two seconds too late, his retreat is cut off, and the powerful jaws of the old dog close upon him.

The next season Cuff tried the same tactics again with like success; but when the third woodchuck had taken up his abode at the fatal hole, the old churner's wits and strength had begun to fail him, and he was baffled in each attempt to capture the animal.

The woodchuck always burrows on a side-hill. This enables him to guard against being drowned out, by making the termination of the hole higher than the entrance. He digs in slantingly for about two or three feet, then makes a sharp upward turn and keeps nearly parallel with the surface of the ground for a distance of eight or ten feet farther, according to the grade. Here he makes his nest and passes the winter, holing up in October or November and coming out again in April. This is a long sleep, and is rendered possible only by the amount of fat with which the system has become stored during the summer. The fire of life still burns, but very faintly and slowly, as with the draughts all closed and the ashes heaped up. Respiration is continued, but at longer intervals, and all the vital processes are nearly at a standstill. Dig one out during hibernation (Audubon did so), and you find it a mere inanimate ball, that suffers itself to be moved and rolled about without showing signs of awakening. But bring it in by the fire, and it presently unrolls and opens its eyes, and crawls feebly about, and if left to itself will seek some dark hole or corner, roll itself up again, and resume its former condition.